Jean-Luc Bouchot

Structures and irregularities in image processing and analysis

Jean-Luc Bouchot

Structures and irregularities in image processing and analysis

Discrepancy norm, distance transform and analytic representations

Südwestdeutscher Verlag für Hochschulschriften

Impressum / Imprint
Bibliografische Information der Deutschen Nationalbibliothek: Die Deutsche Nationalbibliothek verzeichnet diese Publikation in der Deutschen Nationalbibliografie; detaillierte bibliografische Daten sind im Internet über http://dnb.d-nb.de abrufbar.
Alle in diesem Buch genannten Marken und Produktnamen unterliegen warenzeichen-, marken- oder patentrechtlichem Schutz bzw. sind Warenzeichen oder eingetragene Warenzeichen der jeweiligen Inhaber. Die Wiedergabe von Marken, Produktnamen, Gebrauchsnamen, Handelsnamen, Warenbezeichnungen u.s.w. in diesem Werk berechtigt auch ohne besondere Kennzeichnung nicht zu der Annahme, dass solche Namen im Sinne der Warenzeichen- und Markenschutzgesetzgebung als frei zu betrachten wären und daher von jedermann benutzt werden dürften.

Bibliographic information published by the Deutsche Nationalbibliothek: The Deutsche Nationalbibliothek lists this publication in the Deutsche Nationalbibliografie; detailed bibliographic data are available in the Internet at http://dnb.d-nb.de.
Any brand names and product names mentioned in this book are subject to trademark, brand or patent protection and are trademarks or registered trademarks of their respective holders. The use of brand names, product names, common names, trade names, product descriptions etc. even without a particular marking in this works is in no way to be construed to mean that such names may be regarded as unrestricted in respect of trademark and brand protection legislation and could thus be used by anyone.

Coverbild / Cover image: www.ingimage.com

Verlag / Publisher:
Südwestdeutscher Verlag für Hochschulschriften
ist ein Imprint der / is a trademark of
OmniScriptum GmbH & Co. KG
Heinrich-Böcking-Str. 6-8, 66121 Saarbrücken, Deutschland / Germany
Email: info@svh-verlag.de

Herstellung: siehe letzte Seite /
Printed at: see last page
ISBN: 978-3-8381-3662-2

Zugl. / Approved by: Linz, Johanes Kepler University, Diss, 2012

Copyright © 2013 OmniScriptum GmbH & Co. KG
Alle Rechte vorbehalten. / All rights reserved. Saarbrücken 2013

Die Mathematiker sind eine Art Franzosen: Redet man zu ihnen, so übersetzen sie es in ihre Sprache, und dann ist es alsobald ganz etwas anderes.

Maximen und Reflexionen, Nachlaß, Über Natur und Naturwissenschaft

Johann Wolfgang von Goethe

Contents

1	**Introduction**		**1**
1.1	Imaging and images		1
	1.1.1	Imaging process	4
	1.1.2	Image processing and representation	5
	1.1.3	Image analysis and comparison	5
1.2	About the book		6
	1.2.1	Aims	6
	1.2.2	An open window to structure, appearance and singularities	7
	1.2.3	Contributions and organisation	8

I General Background 9

2	**A review of structural similarities**		**11**
2.1	Analysis on the pixel level		12
	2.1.1	SSIM: Structural SIMilarity index	12
	2.1.2	Normalised cross correlation	16
2.2	Variational decompositions		17
	2.2.1	Structure - Texture decompositions	17
		2.2.1.1 The decomposition of Rudin-Osher-Fatemi	17
		2.2.1.2 The decomposition of Meyer	18
	2.2.2	Structure - Texture - Noise decompositions	19
2.3	Wavelets and their extensions		20
	2.3.1	Wavelet transform and image processing	20
	2.3.2	Extensions of the wavelet transform	20
	2.3.3	Denoising and compressing with wavelets	21
2.4	Phase-based image comparison		23
	2.4.1	Phase as structural component of images	23
	2.4.2	Phase-based applications in image processing	24
2.5	Mathematical morphology		26
	2.5.1	Mathematical morphology on binary images	26
	2.5.2	Extension to sampled functions	28

II Analysis of irregularities with the discrepancy norm — 31

3 Theoretical Aspects of the Discrepancy Norm — 33
- 3.1 A century of mathematical research about the divergence of sequences 33
  - 3.1.1 Introduction 33
  - 3.1.2 Domains of applications 34
- 3.2 The modern approach to n-dimensional signals 34
  - 3.2.1 Similarity and misalignment 34
  - 3.2.2 The one dimensional case 36
    - 3.2.2.1 Definition 36
    - 3.2.2.2 Computational aspects 38
  - 3.2.3 Higher dimensional generalization 39
    - 3.2.3.1 On connected components 39
    - 3.2.3.2 Box 39
    - 3.2.3.3 Infinite box 40
    - 3.2.3.4 As a difference of suprimum and infimum 40
    - 3.2.3.5 Equivalence of the autocorrelation functions 40
- 3.3 Characteristic properties 40
  - 3.3.1 A Lipschitz property 41
  - 3.3.2 Non-monotonicity of misalignment functions 41
    - 3.3.2.1 A construction principle 41
    - 3.3.2.2 A non-monotonicity property 42
    - 3.3.2.3 The case of f-divergence metrics 44

4 Discrepancy correlation optimization — 47
- 4.1 One dimensional signals 47
  - 4.1.1 The discrete case 47
  - 4.1.2 Alignment by convolution 49
  - 4.1.3 The continuous case 50
  - 4.1.4 Derivation of the autocorrelation function 54
- 4.2 Higher-dimensional signals 57
  - 4.2.1 Approximation formula 57
  - 4.2.2 Derivation of the correlation function 62
- 4.3 Practical considerations 63

III Other approaches towards structures and irregularities in images — 65

5 Distance transform methods in image processing — 67
- 5.1 Distance transforms for image processing 67
  - 5.1.1 Distance transform on binary images 67
    - 5.1.1.1 Motivation: comparison of edge detection algorithms . 67
    - 5.1.1.2 Binary images as sets 68
    - 5.1.1.3 The distance transform 68

		5.1.1.4 Implementation details	69
		5.1.1.5 Hausdorff metric	70
		5.1.1.6 Image comparaison with distance transforms	71
	5.1.2	Extensions to grey level images	71
		5.1.2.1 Distance transforms along a path	71
		5.1.2.2 Wilson's approach	72
		5.1.2.3 Molchanov's approach	74
		5.1.2.4 Coquin's approach	75
		5.1.2.5 Comparing grey-level distance transforms	76
5.2	Local dissimilarity maps		77
5.3	Multiscale analysis and distance transforms		77
	5.3.1	Scale based edge detection	78
		5.3.1.1 Basic scale-space theory	78
		5.3.1.2 Canny edge detector	80
		5.3.1.3 Comparison of scale dependent edge detections	81
	5.3.2	Multiscale distance transform	81
		5.3.2.1 Continuous distance transform on multiscale edges	81
		5.3.2.2 Discrete distance transform on multiscale edges	83
		5.3.2.3 Some examples	84

6 An analytic approach to structures 91

6.1	Elements of the theory of analytic signals		91
	6.1.1	The Hilbert transform	91
		6.1.1.1 Fourier transforms	91
		6.1.1.2 Hilbert transforms	92
	6.1.2	The analytic signal representation	94
	6.1.3	Properties of the Hilbert transform and analytic signal representation	94
	6.1.4	Analytic signal analysis	95
	6.1.5	The analytic signal as a boundary value problem in complex analysis	97
6.2	Generalizations to higher dimensions		98
	6.2.1	The higher dimensional analytic signal	98
		6.2.1.1 A single orthant description	98
		6.2.1.2 Multidimensional analytic signal analysis	100
	6.2.2	The monogenic signal	101
		6.2.2.1 A boundary value problem in Clifford analysis	101
		6.2.2.2 The Riesz transform and its property	102
		6.2.2.3 Signal analysis with the monogenic representation	104
6.3	Analysis of local monogenic features		105
	6.3.1	Motivation	106
	6.3.2	The not trivial case	108
	6.3.3	The higher dimensional case	110
	6.3.4	Interpretation of phase and amplitude	111
		6.3.4.1 Qualitative results	111

		6.3.4.2	Quantitative analysis	112
	6.4	Comparing higher dimensional analytic signals		113

IV Applications 117

7 Defect detection in regularly textured patterns 119
7.1 Problems and Restrictions 119
7.1.1 Problem statement 119
7.1.2 Restrictions 120
7.1.3 Motivations 120
7.2 A Template-Matching Algorithm 121
7.2.1 The original idea 121
7.2.2 Improvements 121
7.3 Experiments - Results 123

8 Discussion 127
8.1 Conclusion 127
8.2 Opennings 128

List of Figures

1.1 A flowchart of a typical image processing and analysis application. 1
1.2 2
1.3 2
1.4 2
1.5 4
1.6 6

2.1 Flowdiagram of the SSIM index. 12
2.2 Diverse local values of both the original Lena image and a compressed version. We clearly see the importance of the structure in the SSIM index. 14
2.3 A similar example as above but with an image with less textural details. Once again the structures are clearly appearing at each stage of the SSIM computation. 14
2.4 Decomposition of the House image using the ROF algorithm. First row shows the structural component and second one the residual (combination of texture and noise). From left to right the parameter λ equals 1, 5, 10, 15 and 20. 18
2.5 ROF decomposition in the presence of noise. First row shows the noisy image, second one the structure component and third one the textural component. The noise increases from left to right. 19
2.6 Wavelet synthesis after hard (top row) and soft (bottom row) thresholdings with thresholds values (from left to right) of respectively $0.02, 0.06, 0.14, 0, 24$ and 0.40. 22
2.7 Wavelet synthesis after hard (top row) and soft (bottom row) thresholdings in the presence of additive gaussian noise (images on Fig. 2.5). The threshold is proportional to the noise variance. 23
2.8 25
2.9 Effect of amplitude modification in the Fourier domain of an image. In the first row, the Lena image is corrupted with a blob-like background. In the second one, Gaussian noise is added to the original image. 26
2.10 Example of binary image (left) and structuring element (right). 27
2.11 Examples of binary morphological operations. From left to right: Erosion, Dilation, Opening, Closing 27
2.12 Basic morphological operations on gray-valued images. First row makes use of a small disk-shaped structuring elements while the two following ones use rectangle structuring elements with in the order of the size of a brick or of a window respectively. Columns left to right show the dilation, erosion, opening and closing operations. 29

LIST OF FIGURES

2.13 Morphological gradient and Laplacian using a disk shaped structuring element (first two columns) and a small rectangle structuring element (last two columns). 29

3.1 Examples of admissible sets for the definition of the discrepancy norm. 39

4.1 Our dataset of 1 dimensional toy functions used as illustrative examples. 53
4.2 Convergence of the Γ_p functions towards the discrepancy norm. The horizontal axis corresponds to changing values of p while the vertical one is the output of the Γ_p and discrepancy norm functions. 53
4.3 Computed values of the p^* estimator. The continuous curves show the estimated p^* while the the dotted straight line corresponds to the $\overline{p^*}$ as described in the Table 4.1 . 54
4.4 This figure shows how a gradient descent based optimization algorithm might fail when trying to locally optimize the discrepancy correlation function. Indeed the correlation function (second figure) shows some plateau where the derivative (illustrated on the third figure) is 0. On the other hand, using an appropriate p−norm approximation allows to overcome this effect while keeping a really close objective function. 56
4.5 Sample sine wave with their discrepancy autocorrelation functions and their approximate derivatives. The first row shows the input signal extended with 0 padding (first column) and mirroring the wave (2nd and 3rd columns). 2nd row shows the autocorrelation function based on the discrepancy norm and its approximation. Last row shows the approximated derivative together with the finite difference of the discrepancy autocorrelation. 57
4.6 Our dataset of 2 dimensional toy functions used in our paper. 61
4.7 Evolution of the values of p_0 and q_0 depending on the choice of λ. 62
4.8 Evolution of the values of p_0 and q_0 depending on the choice of ε. 62

5.1 Examples of two completely different binary images which have the same FOM score when compared to a same third one. 68
5.2 Influence of the different saturation and normalization parameters of Wilson's Distance Transform. Each row fixes a c (1,2,5,10,100, from top to bottom) while each column fixes a ρ parameter (0.5,1,2,8,16 from left to right). The upper left corner shows the signal used for the transformation. The reference signal is actually of size 256 and ranges from about -2 to 8. 74
5.3 Molchanov distance transform on the House image. First row show the results with varying number of discretisation steps of the intensities while second row show varying value of thresolding parameter. 75
5.4 Molchanov distance transform with different thresholdings of the distances. 76
5.5 Effect of the different edge detections algorithms introduced given at different scales together with their distance transforms. The two first rows involve the Canny edge detector while the two last ones concerns the Gaussian scale space. First and third row are edges at different scales and second and last one are their distance transform (to the foreground). Scale increases from left to right. 82
5.6 Example of images from three databases corrupted only by scaling. We refer to them as scaling sequences 1, 2 and 3. 85

LIST OF FIGURES

5.7	....	85
5.8	Example of images from three databases corrupted only by translation in the camera plane. They will be refered to as the translation sequences 1,2,3 and 4 respectively.	86
5.9	....	87
5.10	Example of images from three databases corrupted only by illumination changes. We refer to them as illumination sequences 1, 2, and 3.	88
5.11	....	89
6.1	An example of amplitude and frequency modulated signal	96
6.2	Original and recovered amplitude and phase of an AM-FM signal.	97
6.3	Spherical representation of the monogenic signal and its local features	105
6.4	A graphical representation of the phase congruency as suggested by Peter Kovesi [Kov99]. The blue arc corresponds to the phase congruency while the black ones corresponds to the phase of different log-Gabor wavelets.	113
6.5	....	114
6.6	....	115
6.7	....	115
6.8	....	116
7.1	Some examples of SIFT keypoints detection on defective image of different textures. The first row shows the gray-level images while the second row shows the corresponding SIFT keypoints found.	120
7.2	Defect detection in textile and airbag hose production after a threshold on the computed LDMap with either the L_2 or the discrepancy norm. Threshold of DN images is 0.8 and of L_2 0.7, with normalized values between 0 and 1.	122
7.3	Examples of the applicability test on textile defect images. For each example the test image (T) as well as the output (O) of the algorithm is shown. Reference images are not shown. The algorithm output is for illustration purposes not thresholded.	124
7.4	Results obtained by a OC-SVM applied with a Gaussian kernel directly on the gray-level values.	125

List of Tables

3.1 Some counter examples for the monotonicity of the gaussian kernel autocorrelation function 36
3.2 Examples of kernels and distance measures that follow the construction principles of Theorem 6 or Corollary 1 with summation as aggregation function 44
3.3 Some examples of f-divergence measures 45

4.1 Examples of approximation and lower bound estimations on some toy functions. . . . 54
4.2 Examples of approximation and lower bound estimations on some toy functions. . . . 62

7.1 Performance comparison of texture characterization approaches using Percentage of Correct Detection (PCD), for details see Tolba et al. [TKMA10]. The original Table contains multiple entries per reference, here only the best performing ones are listed. Furthermore the top performing method of Murino et al. [MBR04] is skipped because it is a pure classification algorithm without detection. The discrepancy norm based algorithm can be compared with the class of Grey-Level Co-occurrence Matrices (GLCM) and filter (Gabor, wavelet) based feature extraction algorithms. 125

$\mathbb{R}$	Set of all real numbers
$\mathbb{Z}$	Set of all, positive or negative, integers
$\mathbb{N}$	Set of all natural numbers
$C\ell_{p,q}$	The Clifford algebra with parameters p and q
$\mathbb{H}$	The set of quaternions
x, y, t	Spatial or time variables
ξ	Variables in the Fourier domain
n	Dimension of the space
M, N, W	Sizes in pixel or number of points
$\Re(z)$	The real part of a complex number
$\Im(z)$	The imaginary part of a complex number
μ	A finite measure on a σ-algebra
F	Cumulative function of the function f
$\mathcal{BV}$	Set of functions of bounded variations
$\mathcal{H}, \mathcal{H}_i, \mathcal{H}_T$	The different Hilbert transforms
$\mathcal{R}, \mathcal{R}_i$	Riesz operators
$\mathcal{F}$	Fourier transform
$\mathcal{QF}$	Fourier transform of quaternionic functions
D_δ	Dilation of parameter δ
τ_t	Translation of parameter t
R_ρ	Rotation of parameter ρ
∂_x, ∂_y	First and second partial derivatives
D^α	Differential operator of order α
$\|\cdot\|_p$	Minkowski norms
$\|\cdot\|_*$	G-Norm of Meyer
$\|\cdot\|_D, \|\cdot\|_I^{(n)}$	Discrepancy norm in one or n dimensions
σ_n	Noise parameter
μ_f, μ_g	Average of the functions f, g
σ_f, σ_g	Standard deviation of the functions
δ	Dirac δ distribution
$\mathbf{1}_A$	Indicator function of the set A
Δ_d	Correlation function based on the (dis-)similarity measure d
M_Φ	Generalised mean
Γ_p	Discrepancy approximation in the one dimensional case
$\Gamma_{p,q}$	Discrepancy approximation in the higher dimensional case
A_f	Set corresponding to the binary function f
$\|\cdot\|$	Cardinality of a set or absolute value
$H(A, B)$	Symmetrical Hausdorff distance between the sets A and B
Γ_f	Subgraph of a function
$X_y(f)$	Upper-level set of a function
d_W, d_M	Wilson's and Molchanov's distances
$\langle \cdot, \cdot \rangle$	Scalar product in L^2

Chapter 1

Introduction

This first chapter of the work is intended as some very coarse background and organisation about the rest of the work. This will help the reader understand what is being done, when, and where. Some interesting definitions are given as well as some details regarding the work done in the last three years. We try to keep the technical details to its lowest here.

1.1 Imaging and images

As this book is about image processing and analysis, we will first give a review of what should be understood by these words. As it can be seen from Fig. 1.1, the image processing and analysis parts are just links in the chain of automatically and numerically understanding the world around us, which can be better described by computer vision.

As images of all sorts are gathered everywhere, the amount of data to be treated is getting huge and the need of interesting algorithms is now mandatory. However, as we will see, the holy grail of image understanding has not yet been found and this work aims at giving ideas on how to work differently with images. For this we should distinguish two main categories of images: the one which represent the world as our eyes see it (we will mainly refer to those ones as *natural image scenes*, see Fig. 1.2 for some classical examples) and the ones which are just representations of physical phenomena, as it

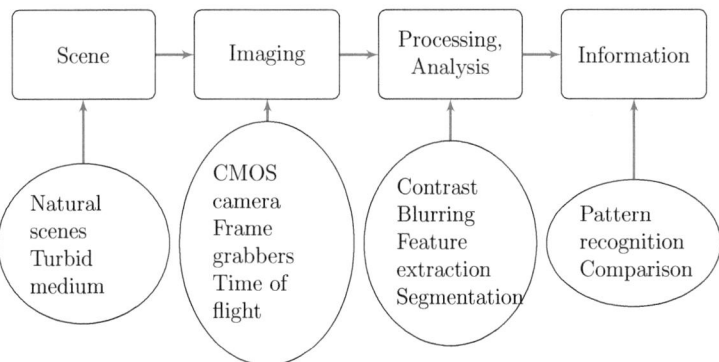

Figure 1.1: A flowchart of a typical image processing and analysis application.

1

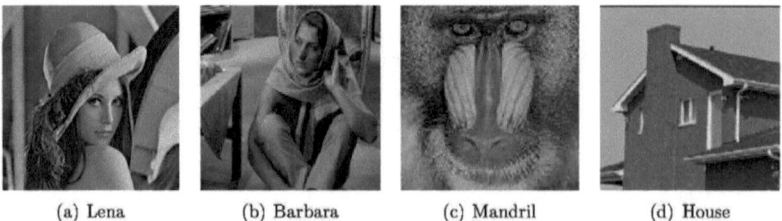

(a) Lena (b) Barbara (c) Mandril (d) House

Figure 1.2: Examples of typical natural image scenes. We will make use of these images as examples for some algorithms throughout this book.

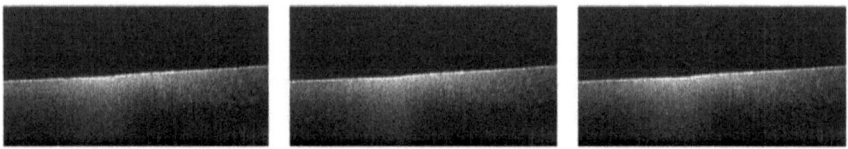

Figure 1.3: Examples of some optical coherence tomography scans.

is the case with Magnetic Resonance Imaging (MRI) or Optical Coherence Tomography (OCT) (see Fig. 1.3 for some examples of polymer material scanned with an OCT imaging).

Unfortunately algorithms developed for one or the other kind of images will not necessarily work well on the other. For instance, if one tries to do feature extraction on OCT images the same way as for natural images, there are reasons to think that it will not work as well as expected. As an example, if one looks at the image displayed in Fig. 1.4, one can see that the features detected by the SIFT [Low04] algorithm are not consistent from one scan to another. This is probably due to the randomness of the speckles which makes the feature points non reproducible.

Even in what we can think would be rather simple image, the SIFT methods might show some weakness. This will be a motivation for investigating a novel algorithm for quality control in regularly textured surfaces (see Chap. 7 and Fig. 7.1)

We will come back briefly on the topic of imaging process, image processing and image analysis in the three next sections. Before starting the more technical details let us motivate the need of such research from both a scientist and a user point of view.

Image understanding, as depicted in the chain of Fig 1.1 relies on many different technical points. Assuming the imaging process is done (which is not necessarily an easy task) several actors take part to this adventure. Mathematicians for instance have to deal with:

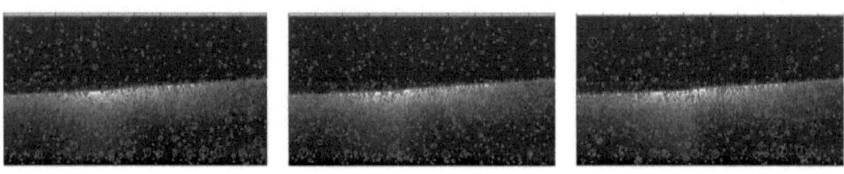

Figure 1.4: Results of the SIFT localisation and description on the OCT images depicted in Fig. 1.3

1.1. IMAGING AND IMAGES

- Numerical approximation of continuous methods: for instance how one can implement the continuous wavelet transform as a discrete wavelet transform; or how one comes to the discrete Fourier transform from the classical Fourier transform.

- Well-posedness of problems: this means analyzing how the final solution depends on the input data. This analysis is particularly interesting when considering noise in the measurements.

and they apply methods coming from:

- Inverse problem; or, given noisy observations, how can I reconstruct the original noise free data,

- Functional analysis; it is used for some functional decomposition or for the description of different function spaces (as used for instance in wavelet analysis),

- Measure and distribution theory; this could for instance be interesting when working with periodic functions in which case we might want to consider working on a torus instead of $\mathbb{R}^n$.

On the other hand, computer scientists are concerned with:

- Parallel programming; which is used when one has access to a cluster of computers,

- Graphical programming; mainly used for single machine with good graphical capacity,

- Embedded CPUs; as an example, we can cite any automatic rovers in space research,

- Real-time systems; as they appear for instance in ultrasound imaging during pregnancy.

While the first ones try to prove the feasibility of a given algorithms the following ones are concerned with its applicability to real-world problems. But both actors could not work without each other and actually most of us are working at the frontier.

For the automatic understanding one should also not forget the progress made in the area of machine learning which lies also at the border between applied mathematics (with integral operators and the theory of Hilbert spaces for kernel machines for instance) and computer science (e.g. for the numerically stable solutions of large scale learning algorithms).

Finally as last but not least actors in the image understanding process are the psychologists and neuroscientists. Indeed one of the most promising and successful algorithms ever done in this area (the neural networks) is based on a numerical modelisation and simulation of the human brain.

The image processing framework as presented above is used in many different ways for many different applications. One can cite the latest barcode reader on mobile device to get information about concerts for instance. The latest smartphones can also be used as a terminal for visiting museums by using augmented reality. Image understanding finds also application in the security domain where face or finger print recognition might be needed. More lately, in the context of human gesture analysis, videos are analysed in order to detect odd behaviors in crowded environments. Image understanding is also present in the industrial world where one wants to lower the cost of quality control while keepingthe accuracy of a human controller (often better due to the non-tiredness of the machine). Finally, lately image understanding as been used for entertainment purposes as used in video games.

1.1.1 Imaging process

There are different ways to get an image of the world. The most common one is used for instance during vacation to keep an image of the world as we see it: colours, objects, and people appear. For this purpose small sensors collect the light intensity emitted, or actually reflected, by the different objects in the environment. This yields *natural image scenes*. However, we can think of other physical to be measured and in that case, the images obtained might look completely different than what one sees through their eyes. As an example, we can cite the Laser imaging technologies or the Time of Flight cameras [GYB04, STDT08] for depth measurements.

While in the first example direct measurement of the intensity of a wave is done on the sensor and this information is saved and eventually displayed. In the second case, two time stamps are saved: one when the Laser beam is sent and the other one when it comes back to the sensor. The duration of the travel allows to compute the distance between the camera and the reflecting object knowing which wavelength was used.

Another approach is when considering interferometric patterns. In this case we no longer measure physical coming from an object directly but rather some interferences between a beam with itself after a small time shift in a tissue sample for instance. Fig. 1.5 shows a spectral domain polarization sensitive OCT *(SD-PS-OCT)* setup where we have a reference mirror (RM), polarizer (P), beamsplitter (BS), polarizing BS (PBS), quarter wave plates (QWP), galvano-scanner mirror (GM), diffraction gratings (DG) and line cameras (CCD) .

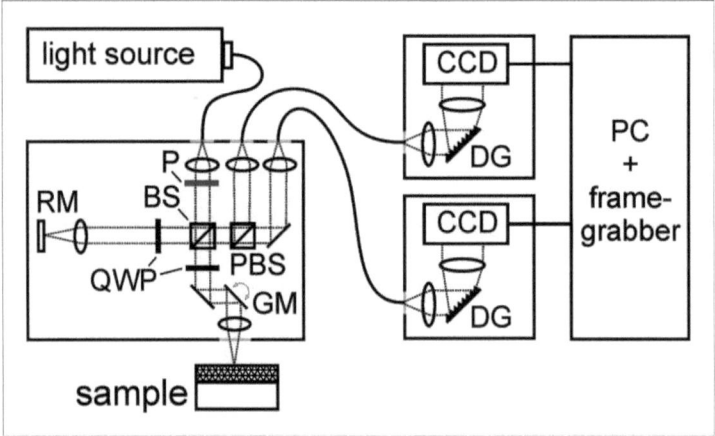

Figure 1.5: Example of interferometer used in OCT imaging for material sciences. Initials are explained in text.

The importance here is that the beam is being split and while one part of it is not changed the one going through the sample will encounter a difference in the optical path length inducing some interferences, as long as the coherency is kept. More information and details about this procedure can be found in the literature [HHSF92, dBMvGN97]. Such images have drastically different characteristics than the natural scenes and particular care should be taken when analysing them [SLHH[+]11, LHCH[+]12]

1.1.2 Image processing and representation

Note that some people do not make the difference between image processing and image analysis. However, as we will develop along this work, we make here a big difference in the following sense: an image processing algorithm changes the representation or enhance the quality of an image. It means that an image processing procedure will take an image as an input and gives back another image. For instance, when one applies the Fourier transform to an image, the information entailed in it is not change but instead of having intensity information, one gets information about the frequency repartition in the image.

On the other side, image analysis is more concerned with extracting information or comparing images. However this separation is both only the author's view and not crisp. Indeed the process of extracting edges can be seen as an image analysis for some people as the information is being strongly modified but we will consider it as a processing step. Indeed, as we will see later in the book, we will use the edge detection procedure only as a mandatory step before comparison.

We introduce along this study only digital image processing algorithms but these can also be done directly on the optics, for instance when using Spatial Light Modulator (*SLM*) [HSM+12].

We have already cited the Fourier transform and the wavelet transform. Both algorithms are examples of image processing tools. One could also note any kind of low level image transformation such as blurring and deblurring, contrast enhancement or γ-correction. We will develop later other representation which are based on distance transforms or analytic/monogenic representations.

1.1.3 Image analysis and comparison

Once again, the frontier is not really clear. Whether image analysis should be understood as high-level computer vision, with methods such as feature extractions, or more low-level image comparison is not clear. However within this analysis, we stick to the case of comparison at the pixel level. However, this pixel might have been transformed by some previous image processing operations. In particular, if one considers the case of monogenic image analysis [BBRH13], we see that we no longer deal with intensities but other physical values are represented.

We can also consider a big part of this research area as overlapping with the theory of similarity measures. While similarity measures can be applied to anything, we consider here only image comparison and try to derive suitable similarities, for instance, for registration purposes or for recognition purposes. However, a similarity measure should be developed with a proper application in mind.

Consider for instance the images depicted on Fig. 1.6 and assume in addition that we are given a similarity measure s with values in $[0, 1]$, where a 1 occurs if and only if we have a perfect match. The images are all obtained by some translations of an original one. In the case of a recognition problem, we would expect the similarity measure to yield values close to 1. However, if we are interested in alignment, then the similarity measure should reflect the displacement, and, if possible, in a monotonic way (see for instance the image on the right hand side of Fig. 1.6). For this task, we can have a look at the discrepancy norm [Mos11] which is well suited when analysing misaligned patterns.

Fig. 1.6 shows a sample function of two white bars being translated to the left on a dark background. while for a recognition task, we should get a similarity measure close to 1 almost every time, the last image shows how a dissimilarity should behave in an optimal manner against shifts for alignment tasks.

Figure 1.6: Examples of typical natural image scenes. We will make use of these images as examples for some algorithms throughout this book.

1.2 About the book

This section is more concerned with the roots, objectives and organisation of this work.

1.2.1 Aims

The aim of this study to give the reaser some understanding of both the theoretical and practical aspects of the structures in images.

In the work of Moser [Mos11], it has been shown that an autocorrelation function based on the discrepancy norm shows some monotonicity property with respect to shifts (Note that more information on that topic is be given in Chap. 3). It therefore makes sense to study its behaviour in the context of (eventually multi-dimensional) signal processing and misalignment functions.

Some tasks were thought of as eventually suitable for the proposed norm and had to be studied within the project by the two students. We were mainly concerned on the two dimensional signal processing as it directly applies to images. We were therefore particularly concerned with computer vision as well as image processing. These are 2 different concepts even if a direct connection exists. For instance when considering the process of fusing two images (*i.e.* merging information of a same object of phenomenon with different imaging methods or representations), which is conceptually a computer vision task, a segmentation of both images onto region of interests, which is conceptually an image processing task, is often a first step for the alignment.

However, when we already know exactly which object we are trying to segment, we can improve the alignment and conversely, when we have a nice alignment of both images, we can improve the segmentation process and so on. This chicken and egg problem makes it hard to generalize all the algorithms, and therefore both theories (image processing and computer vision) should be studied with care.

In a more scholar way, our homework was to study the applicability of the discrepancy norm in the (applied) context of:

- correlation analysis in time-series
- registration

Some more theoretical points had to be studied too:

- construction principle of similarity measures
- fast computation

Time series analysis In this part we wanted to detect certain patterns within fast varying functions. This is done by applying a discrepancy norm as well as a faster variant which can be computed by convolution.

1.2. ABOUT THE BOOK

Registration Registration is the problem of aligning two given patterns where one is distorted by a certain transformation from the other one. Thanks to the monotonicity property of the autocorrelation function of the discrepancy norm, our similarity measure is particularly suited for such tasks. We see how it can be applied for pattern alignment as well as some industrial examples of quality control.

Constructing similarity measures As a side effect of the research about the discrepancy norm, we have studied this remarkable property that an increase in the misalignment entails an increase of the discrepancy autocorrelation function. As it turns out, only few similarity measures can fulfil this property and a construction principle has been introduced to develop this idea.

Approximation - Fast Computation A big part of the work has been spent on accelerating the computation of the discrepancy norm as well as improving its differentiability. We will see in a later sectino that the discrepancy norm shows some strongly non-differentiable point in its autocorrelation function as well as some plateau (defined as regions with vanishing gradient) which yield harder optimisation procedure. Therefore general formulae for the approximation of the discrepancy norm have been developed and can be implemented, in the one dimensional case, with two convolutions for solving the alignment problem. Some sensitivity analysis has been done to ensure convergence of the approximation within a certain error bound.

1.2.2 An open window to structure, appearance and singularities

Along the researches about the discrepancy norm and parallel similarity measures, one can notice that some other similarities fulfil a similar monotonicity property. Among others, the Hausdorff distance, used for instance for image recognition purposes, is particularly interesting as it come handy for distance transform algorithms. As a consequence we introduce a novel algorithm as an extension of the distance transform to non-binary images. It is based on a scale space representation and edge detection. With this representation, some ideas for image comparison are given together with a first analysis of their performance.

Now it is reasonnable to wonder why all the concepts presented in the tbooks are gathered under the terms *structures* and *irregularities*. As it is detailed in the chapter about the discrepancy norm, its theory comes from the analysis of the odd, or regular, behaviour of numbers along the unit circle whence the irregularity component of the title. This aspect is also responsible for the structure part: the discrepancy norm tends to consider the pixel or data only on a domain with the highest divergence. In other terms, it tends to emphasize the abrupt changes within a signal or an image. Abrupt changes are actually a way of representing edges, or structures.

Structures and irregularities also appear in the context of distance tranforms as they are directly driven by the edges (for the structural part) or by salt and pepper noise (for the irregularity part). As a last concept introduced and developed in this book, the monogenic and analytic representation are examples of structural image transformations. It indeed tends to increase singularities (in a mathematical sense) in images. In this sense it tends (when analysing the phases) to emphasise the structural component of an image.

1.2.3 Contributions and organisation

Organisation This book is articulated in four main independent parts. The first part deals with some general background about image understanding from the point of view of structures. We first review the state of the art in image processing and analysis where irregularities or structures have a role to play. We have tried to make as complete and interesting as possible but due to the huge amount of literature on that topic we can not claim it to be exhaustive. However most of today's topic of research are coarsely covered in this chapter.

The second part is dedicated to the theoretical study of the discrepancy norm. The first chapter details the already existing theory regarding the discrepancy norm in image and signal analysis and introduce our first result for constructing similarity measures in a general way. A theorem states that such similarities, though very general, cannot fulfil our monotonicity condition for misaligned patterns.

The second chapter introduces a novel approximation of the discrepancy norm. It follows a rather incremental schemes: we first develop the approximation together with the first results for discrete univariate signals and then extend the results first to continuous signals and then to the multivariate case.

The third part summarizes other approaches. A first chapter is dedicated to the study and applicability of distance transform algorithms to image processing and comparison. It reviews the already existing state-of-the-art in the domain and finishes by introducing an image representation based on a multiscale edge detection framework.

The second chapter of this part deals with the theoretical study of analytic and monogenic representations and derive their mathematical motivation as boundary value problems in either multidimensional complex analysis or Clifford analysis. The differences between each representation are illustrated on some toy examples for AM-FM signal demodulation.

The last part shows an application of the discrepancynorm. We mainly develop an algorithm for fault detection in regularly textured surfaces by means of local dissimilarity analysis.

Part I

General Background

Chapter 2

A review of structural similarities

As we are interested in this book in structures and singularities in images, we begin this chapter with a small review of existing methods for analysing or decomposing images into different information representation. A common characteristic of all the representation and analysis we are going to present here is the fact that they, somehow, give a crucial role to the structural component of an image. While we do not claim to be exhaustive, we hope to give a rather global overview of what is being done actually.

Unfortunately, there is no uniform definition of the structure of an image, and one should be careful with such expression. However, we can try to give a general qualitative explanation: a structure can be understood as any abrupt change in the image intensity, at a certain scale.

We concentrate on pixel based measures. In order to give a complete state of the art in image representation and comparison, we should describe (among others) feature based methods and bag of words representations. However, this topic would drive us far away from the main idea as the researches about feature extraction and descriptor computations are very fruitful. For the curious reader we refer to [NA08] for some interesting materials.

In this chapter, we first give some details about two structural image comparison metrics which do not need any optimisation or change of representation: the Structural SIMilarity [WBSS04] index and the Normalised Cross Correlation [Lew95].

Then we will have a look at variational method for image decomposition. The basic idea relies on the fact that an image can be understood as a superposition of different layer. The most known and original decomposition [ROF92] is of the form $f = u + v$ where u represents a structural component and v a textural one.

As another important category of structural analysis and representation method, we recall the work based on Fourier spectrum phase. These techniques are well known since the work of Oppenheim [OL81] where the author has shown the importance of phase in images. The other reason to look at the Fourier spectrum is because of the Fourier translation property stating that translation in the space are converted into phase shifts in the Fourier domain, making it useful for instance in the context of image registration.

We conclude this chapter by giving some details regarding mathematical morphology.

2.1 Analysis on the pixel level

2.1.1 SSIM: Structural SIMilarity index

Introduction The Structural SIMilarity index (*SSIM*) [WBSS04] was introduced as an alternative method to the root mean square error or the peak signal to noise ratio (*PSNR*). It has been mathematically studied only very recently [BVW12]. This method works by splitting an intensity image information into three component: two real numbers assessing the overall illumination and contrast of the image, and a normalized image which is said to contain only structural information. Finally these three components computed for both reference image f and measured image g are compared independently and the results of those comparisons are combined to give a similarity score.

The general process for computing the SSIM score is depicted on Fig. 2.1 (taken from [WBSS04])

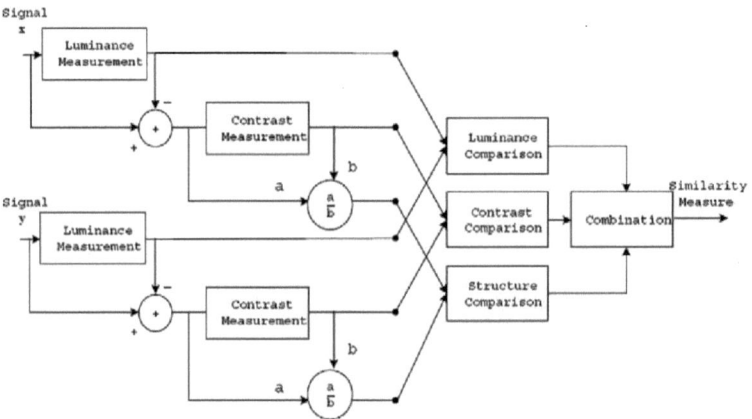

Figure 2.1: Flowdiagram of the SSIM index.

Formally speaking, a reference image is a function f defined on a discrete finite pixel grid with values in (at first) a positive finite subset of $\mathbb{R}$: $f : D \subset \mathbb{Z}^2 \to [0, L] \subset \mathbb{R}$, where L denotes the maximal achievable value. For instance, for an 8 bit encoded gray scale image, we have $L = 255$.

The luminance (or *intensity*) information is first computed by calculating the average intensity over the whole image $\mu_f = \frac{1}{|D|} \sum_{x \in D} f(x)$. The contrast information is computed as the variance of the intensity: $\sigma_f = \left(\frac{1}{|D|-1} \sum_{x \in D} (f(x) - \mu_f)^2 \right)^{1/2}$. Finally the original image is normalised with respect to both illumination and contrast information and the authors claim that only the structural information, which have zero mean and unit variance, are left: $\widetilde{f} = \frac{f - \mu_f}{\sigma_f}$

So in order to assess the quality of a distorted image g given a reference one f, we need to compare each of the three illumination, contrast and structural information. The authors proposed to make use of the following framework:

$$SSIM(f, g) = \mathfrak{A}\left(l(f, g), c(f, g), s(f, g)\right) \qquad (2.1)$$

where $\mathfrak{A}$ corresponds to an aggregation function in charge of combining all three kinds of information

2.1. ANALYSIS ON THE PIXEL LEVEL

together, l, c and s are the **l**uminance, **c**ontrast and **s**tructural comparisons.

The authors suggest then to use the followings:

$$l(f,g) = \frac{2\mu_f \mu_g + C_1}{\mu_f^2 + \mu_g^2 + C_1} \quad (2.2)$$

$$c(f,g) = \frac{2\sigma_f \sigma_g + C_2}{\sigma_f^2 + \sigma_g^2 + C_2} \quad (2.3)$$

$$s(f,g) = \frac{\sigma_{fg} + C_3}{\sigma_f \sigma_g + C_3} \quad (2.4)$$

$$\mathfrak{A}(l,c,s) = l^\alpha c^\beta s^\gamma \quad (2.5)$$

C_1, C_2 and C_3 are chosen as a small fraction of the dynamic range L to avoid numerical instability. The three powers α, β, γ are fitted to give different importance to the different components. The structural comparison corresponds to a correlation coefficient of the normalised images.

For the special case of $C_3 = C_2/2$ and when all three components have the same importance (*i.e.* $\alpha = \beta = \gamma = 1$), we get the following score, which is the one used by the authors:

$$SSIM(f,g) = \frac{(2\mu_f \mu_g + C_1)(2\sigma_{fg} + C_2)}{\left(\mu_f^2 + \mu_g^2 + C_1\right)\left(\sigma_f^2 + \sigma_g^2 + C_2\right)} \quad (2.6)$$

This global image comparison however does not get localised information of the image but rather tends to average all neighborhood over the pixel grid. It is therefore recommended to compute such features locally by defining for instance a Gaussian neighbourhood around each point of the pixel grid. This yields a dissimilarity map which actually contains well localised luminance, contrast and structural information which we denote by $SSIM^w$ (w standing for the kind of neighbourhood chosen).

As for the rest of the experiments we use a Gaussian weighting function on an 8×8 neighbourhood. We now need a coordinate-wise aggregation function to get a global dissimilarity measure. For this, we stick to the authors' choice of using a classical average and define

$$MSSIM(f,g) := \frac{1}{|D|} \sum_{x \in D} SSIM^w(f,g)(x) \quad (2.7)$$

The local values as well as the dissimilarity maps are displayed for two examples in Figs. 2.2 and 2.3.

In these figures two images (the Lena image 2.2(a) and the house image 2.3(a)) are compressed with a JPEG algorithm [SCE01]. Their compressed versions are compared to the original images using the SSIM framework. Both figures show the local averages as well as the standard deviations and the structural components which are left after normalisation. Moreover the last column shows the dissimilarity maps as well as the structural differences.

As it can be seen, mainly from the house image, the structural component has a strong effect on the SSIM map. For instance the brick-like texture is being deleted by the compression algorithm and as a consequence plays an important role in the dissimilarity maps on the last column. The choice of the compression as a distortion used for illustration purposes comes from the fact that it is known that it produces artifacts creating superfluous structures.

We now give details on some improvements or extensions made to the Structural SIMilarity index in order to cope better with translation or dissimilarities at different scales.

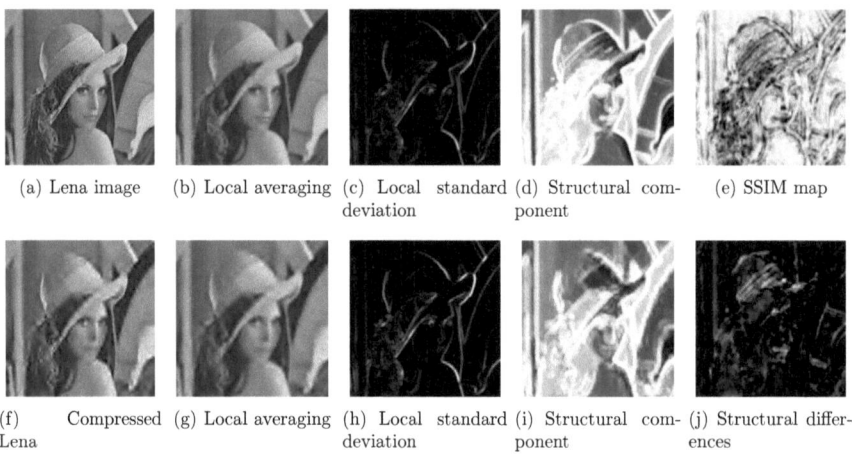

Figure 2.2: Diverse local values of both the original Lena image and a compressed version. We clearly see the importance of the structure in the SSIM index.

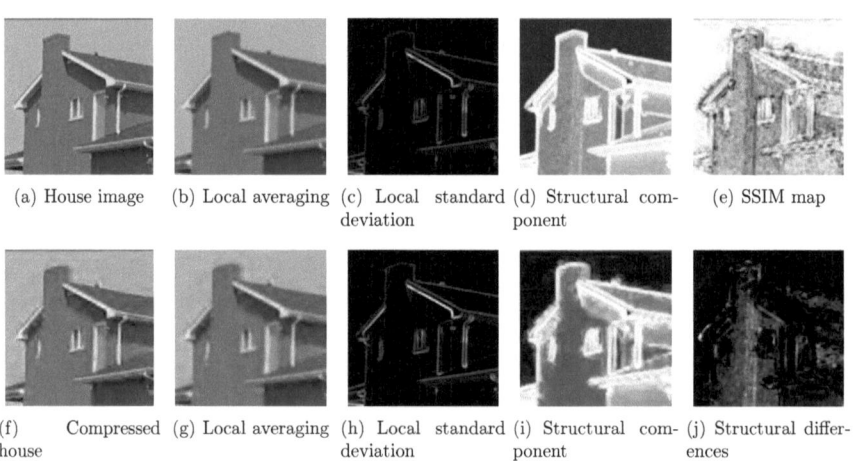

Figure 2.3: A similar example as above but with an image with less textural details. Once again the structures are clearly appearing at each stage of the SSIM computation.

2.1. ANALYSIS ON THE PIXEL LEVEL

Extensions Since this novel work, researchers have been showing growing interest regarding usefulness. While we do not give too many details, we try to give the reader a certain overview of what is being done actually.

The first interesting development was the introduction of multiscale analysis [WSB03] for the SSIM index. It works as the original one except that the contrast and structural information are computed for M different scale, while the luminance is computed only once for the whole image. The multiscale representation is done by simply applying a low-pass filter and downsampling the image by a factor of 2. The quality score finally reads

$$MS-SSIM(f,g)(x) = l^{\alpha_M}(x) \prod_{j=1}^{M} c^{\beta_j}(x) s^{\gamma_j}(x) \qquad (2.8)$$

where the different exponents are chosen to adapt the importance of the global luminance information compared to the structural and contrast information at each scale.

A next improvement [SWG+09] consists in considering an SSIM index after a complex wavelet transform. It was intended to counteract the poor results of the SSIM index in spatial domain against small rigid distortions such as rotation, translation and scaling. While it is important not to have insensitivity to such geometric changes when doing registration, it is arguable when assessing the quality of an image given a reference. Indeed, if one wants to get closer to the human visual systems, such distortion should not affect the quality score the same way as blurring or compression which tend to break the proper structure of the image.

Finally the comparison is done on the complex wavelet coefficients and can be written as a product of magnitude and phase information. The magnitude part of the score has a similar role as the spatial domain SSIM while the phase information should contain most of the local structure (see [OL81] or later in this chapter for more details).

Another variant arises based on the following observation: SSIM does not deal well with blurring. It is meant that small blurring are acceptable for high quality images but somehow, the nature of the image can be strongly affected by high blurring. However, the SSIM does not reflects this (subjective) appreciation. In [CYX06] the authors suggest to compute the SSIM on a gradient map instead than on the intensity map. This yields the following quality score:

$$GSSIM(f,g) = \mathfrak{A}\left(l(f,g), c(\partial f, \partial g), s(\partial f, \partial g)\right) \qquad (2.9)$$

using the notation introduced for the SSIM coefficient. For the derivative computation they used a classical 3×3 Sobel filter. The magnitude of the gradient is computed by taking the biggest component within the two directions.

This idea of using gradient as input to the structural and contrast information is also used to speed up the computation, as in the *Fast-SSIM* [CB11]

A last preoccupation is when we are no longer dealing with mono-channel images but with multi-channel data. In this case, we need a score which takes into account the topology of the colour space. A first tentative was to embed a color image in a hue-saturation-intensity colour space [SDZL09] and to combine a hue-correlation component $H(f,g) = \frac{2h(f)h(g)+C_4}{h(f)^2+h(g)^2+C_4}$ with the classical SSIM measure. This combination is done by a linear combination of both correlations.

A more recent attempt [KYP12] to generalize the SSIM to colour images is based on quaternion representation. In this case, an image is no longer seen as a vector valued function but as a quaternion valued function instead. Then an SSIM-like score is derived by means of calculus on quaternion. It is believed that this approach can:

1. be generalized to any other gray scale visual quality assessment score

2. handle luminance distortion, chrominance distortion and combined luminance-chrominance distortions (which is not the case for other known metrics)

While this last review is not exhaustive, it gives the reader a rather general overview of the state of the art researches in the context of SSIM and its variants.

2.1.2 Normalised cross correlation

The reason for the presence of the normalised cross-correlation in this review is that, as it appears in the definition, due to the normalization process in the calculation, it acts actually as if there were locally no contrast or intensity information, as done in the normalisation process of the SSIM. Given a template t and an input function f, the normalised cross correlation is expressed as

$$NCC(u,v) = \frac{\sum_{(x,y)\in w} \left(f(x-u,y-v) - \overline{f}_{u,v}\right)\left(t(x,y) - \overline{t}\right)}{\{\sum_{(x,y)\in w}\left(f(x-u,y-v) - \mu_f(u,v)\right)^2 \sum_{(x,y)\in w}\left(t(x,y) - \mu_t\right)^2\}^{\frac{1}{2}}} \quad (2.10)$$

where we define w as the window on which the template is defined (for instance, t might be defined on a pixel grid of size $M \times N$), and where $\mu_f(u,v)$ represents the average of the function f taken over an area of size $M \times N$ centered at pixel (u,v); μ_t represents the average illumination of the template.

As we can see in the equation, the cross correlation is actually computed on images having locally zero mean and unit variance; therefore it is expected that the residual images would contain mainly structural information.

Interestingly, since the work of Lewis [Lew95] it is possible to compute this normalised cross correlation efficiently by means of integral images [Cro84].

The idea here is to separate numerator and denominator and noticing that the numerator can be computed by normal cross-correlation and application of the Fourier theorem:

$$\begin{aligned}\sum_{(x,y)\in w}\left(f(x-u,y-v) - \overline{f}_{u,v}\right)\left(t(x,y) - \overline{t}\right) &= \sum_{(x,y)\in w} f(x-u,y-v)\left(t(x,y) - \overline{t}\right) \\ &\quad - \overline{f}_{u,v}\sum_{(x,y)\in w}\left(t(x,y) - \overline{t}\right) \\ &= \sum_{(x,y)\in w} f(x-u,y-v)\left(t(x,y) - \overline{t}\right)\end{aligned}$$

as the second term is a product of a sum with a local zero-mean function.

The numerator is a little more tricky but using the summed of area tables solves the problem:

$$s(u,v) = f(u,v) + s(u-1,v) + s(u,v-1) - s(u-1,v-1)$$
$$s^2(u,v) = f(u,v) + s^2(u-1,v) + s^2(u,v-1) - s^2(u-1,v-1)$$

where s and s^2 denotes the sum of all intensities, respectively squared intensity, left and above the

2.2 Variational decompositions

2.2.1 Structure - Texture decompositions

These ideas come from the pioneer work of Mumford and Shah [MS89] who introduced the cartoon model for images using calculus of variations and energy functionals to minimise.

As an example, the original idea of Rudin et al. [ROF92] was about denoising images. The idea was that an image is always corrupted by some noise which can be modeled in an additive manner. They propose to try to recover the original noise-free image f from the noisy observation f_0 by minimizing a functional, which yields what can be called as a structural information. Note that this denomination is not really correct, as this component does contain some textural information.

However, this was the first example of decomposition in the form $f_0 = f + noise$, where f plays the role of the original *structural* image and $noise = f_0 - f$ represents the noise (or what will later be *textural*) component.

After this first work, which we will detail in the next section appeared many different approaches for noise and structure separation. A similar approach, based on a dual optimisation problem was proposed by Meyer [Mey01, Mey03]

2.2.1.1 The decomposition of Rudin-Osher-Fatemi

We only give here the basic ideas regarding this algorithm. The curious reader should study the literature should read the suggested literature as well as the references within.

A bit of theory In their work the authors try to recover a noisy image f from an noisy observation $f_0 := f + noise$ where $noise$ is assumed to be an additive gaussian white noise with 0 mean and standard deviation σ_n. The idea is to consider $f \in \mathcal{BV}$, a function of bounded variation. In other terms, it total variation norm has to be bounded. The total variation norm is defined as the L^1 norm of the gradient:

$$\|f\|_{BV} := \|\nabla f\|_1 = \int_\Omega |\nabla f| \qquad (2.11)$$

with Ω being a domain of definition of the function f. It is generally assumed to be smooth and bounded. In case of non-smoothness a unique continuation principle can be used, and the function can be extended on a better domain.

Note that $|\nabla f|$ can take different forms such as $|\nabla f| = \sqrt{\sum_i f_{x_i}^2}$ (for what is called the $TV - L^2$ scheme) for the isotropic case or $|\nabla f| = \sum_i |f_{x_i}|$ for an anisotropic version (or $TV - L^1$).

The algorithm solves the following optimization problem:

$$f^* = \underset{f \in BV}{\operatorname{argmin}} \left\{ \|f\|_{BV} + \lambda \|f - f_0\|_2 \right\} \qquad (2.12)$$

where λ acts as a regularisation or fidelity parameter. As λ tends to 0 the denoised image tends to be more uniform.

As an example here, we have used the implementation done by Pascal Getreuer [1] which is based on Split-Bregman iterations [GO09]. Other methods might be considered (for instance Chambolle's algorithm [Cha04]).

Total variation methods are an active research field in image processing and applications can be found in denoising [ROF92], inpainting [CL97] or segmentation [CV01]. It also appears in the recent research about compressive sensing [LDP07] or, in another form, in dictionary learning for sparse representation [MBP+08].

Some examples We investigate here some examples of texture and structure decomposition based on the ROF model. While the model was developed for denoising purposes, we can actually assume it to be a decomposition $f = u + v$ with u a structural component and v a textural component [Mey01]. Fig. 2.4 illustrates this behaviour.

Figure 2.4: Decomposition of the House image using the ROF algorithm. First row shows the structural component and second one the residual (combination of texture and noise). From left to right the parameter λ equals 1, 5, 10, 15 and 20.

The next figure (Fig. 2.5) shows a similar decomposition when the input image is corrupted by additive Gaussian noise with different intensities.

As we can see in this last set of images, the noise tends to overcome the textural information in the v part of the $f = u + v$ decomposition. This will actually motivate the choice of a more complex decomposition containing a noise component (see later, Sec. 2.2.2).

2.2.1.2 The decomposition of Meyer

It's the dual approach to the previous one. One tries to reconstruct $v \in \mathcal{BV}'$ a strongly oscillating function. This subsection is intended only as a coarse overview of what is being done in current research and should not be considered as a well founded exhaustive review of such methods. It would need indeed a much deeper mathematical comprehension of G spaces [Mey01] and their associated $\|\cdot\|_*$ norm. The principal difference here is that the textural part is no longer considered as the

---

[1]This implementation is available on his personal website: http://www.getreuer.info/home/tvreg or on the Matlab file exchange server: http://www.mathworks.com/matlabcentral/fileexchange/29743-tvreg-variational-image-restoration-and-segmentation

2.2. VARIATIONAL DECOMPOSITIONS

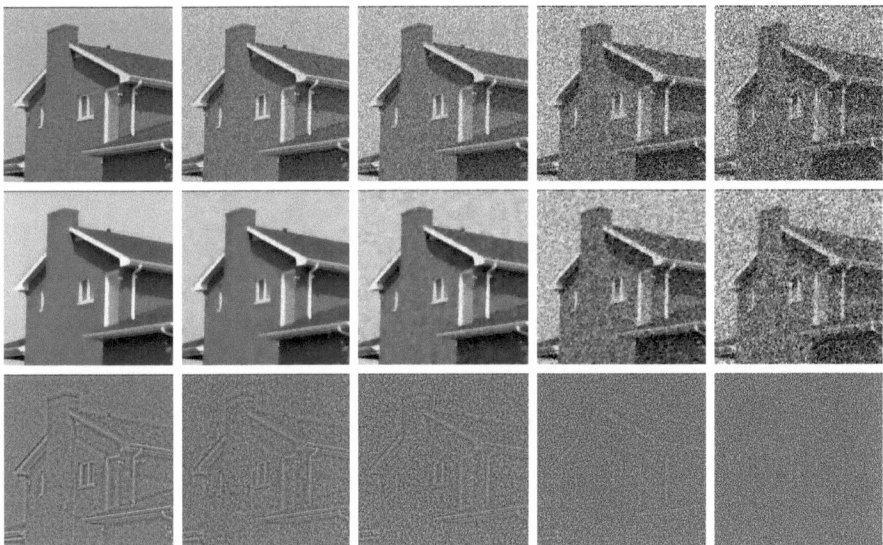

Figure 2.5: ROF decomposition in the presence of noise. First row shows the noisy image, second one the structure component and third one the textural component. The noise increases from left to right.

difference between the observation and the structural component but is rather modeled as a function in this G space. Therefore, Meyer suggests, and so does one of its improvements [HM07], to minimize the following functional

$$J(u) = \|u\|_{BV} + \|v\|_* \qquad (2.13)$$

and the $\|\cdot\|_*$ is defined as

$$\|f\|_* = \inf\left\{\|g\|_\infty = \left\|\sqrt{\sum_i g_{x_i}^2}\right\|_\infty : f = \operatorname{div} g\right\} \qquad (2.14)$$

2.2.2 Structure - Texture - Noise decompositions

A main drawback of the previous decomposition appears in the presence of noise. Noise can indeed be explained as a strongly oscillating function which overcomes the textural part of the image. Gilles studies in his thesis [Gil06] a new decomposition which yields not 2 but 3 image components: $f = u + v + w$ with u the "pure" image and v the *textural component* and w is understood as the noise in the image.

One should notice that the G-norm gives an indication about the oscillating capacity of a function: a purely cosine function of frequency ω will have a G in the order of $1/\omega$ which means that uniform functions ($\omega \to 0$) will have infinite norm while this norm will get smaller for higher frequency oscillating patterns. Having this in mind, it makes sense to look at the noise as an oscillating component having very small G norm, while the textural component, also considered oscillating, shows a smaller frequency.

Therefore the functional to be minimised in this case reads

$$J(u,v,w) := \|u\|_{BV} + \|v/\mu_1\|_* + \|w/\mu_2\|_* + \frac{1}{2\lambda}\|f - u - \nu_1 v - \nu_2 w\|_2 \qquad (2.15)$$

where we have u a function of bounded variations ($u \in \mathcal{BV}$), v a slowly oscillating pattern ($v \in G_{\mu_1} := \{v \in G : \|v\|_* \leq \mu_1\}$) and w a very strongly oscillating pattern ($w \in G_{\mu_2}$). ν_1 and ν_2 act as locally adaptive regularisation coefficients (see for instance [GSZ03] for more information on that topic). To ensure that w oscillates faster than v we must choose $\mu_2 < \mu_1$. Solution can be found [Gil06] based on projections on G spaces.

Remark that this algorithm is not the only one dedicated to a decomposition into three components and Jerome Gilles' thesis is a good source of information regarding such variational decompositions.

2.3 Wavelets and their extensions

This section is intended only to make a rather complete description of the state of the art about structures in images and while the algorithms will not be detailed further, we suggest the reader to study the references given hereafter. It is by no way to be considered a complete study of wavelet methods in image processing.

2.3.1 Wavelet transform and image processing

The wavelet transform was first introduced by Haar [Haa10] and then further studied mainly by Mallat [Mal89, Mal99] and Daubechies [Dau88a, Dau90]. It relies on the following transformation:

$$W_\psi f(s,t) := \frac{1}{\sqrt{s^n}} \int_{\mathbb{R}^n} \overline{\psi\left(\frac{x-t}{s}\right)} f(x) \mathrm{d}x, \qquad (2.16)$$

while ψ denoting what is called the *mother wavelet* and s and t denoting the scale and time or shift parameters respectively. The function $x \to \psi\left(\frac{x-t}{s}\right)$ is often written as $\psi_{s,t}$ in the literature and represents a scaled and translated variant of the mother wavelet. The mother wavelet should fulfil some conditions [Füh96] regarding its mean value (should be 0) and its integrability in the Fourier domain (in one dimension, the integral $\int_{-\infty}^{\infty} \frac{|\widehat{\psi}(\xi)|^2}{|\xi|}$ should be finite). Under these conditions, it can be shown that the Wavelet transform can be used as a frame and we can have a decomposition at discrete scales and space by dyadic decomposition.

2.3.2 Extensions of the wavelet transform

While wavelet analysis shows useful properties for the analysis of one dimensional signals, it however lacks some strength when extended to higher dimensions. It is indeed mainly driven by the direction of the bases of $\mathbb{R}^n$ and does not manage to get the concept of curves. The main reason for that is that their extensions to multidimensional signals are using a separability principle. Because of that they cannot entail the inherent curved structures of an image. To overcome this problem Candès suggested to add an orientation parameter to the wavelet transform [Can98]. This yields the following definition

2.3. WAVELETS AND THEIR EXTENSIONS

Definition 1 (Ridgelets). *A ridgelet can be defined as a directed wavelet on a two-dimensional domain:*

$$\psi_{s,t,\theta} : \mathbb{R}^2 \to \mathbb{R} \tag{2.17}$$

$$(x,y) \mapsto \frac{1}{\sqrt{s}}\psi\left(\frac{x\cos(\theta)+y\sin(\theta)-t}{s}\right) \tag{2.18}$$

Note that this ridgelet is located in space, scale and orientation. Moreover, a similar representation can be given for higher-dimensional signal where we no longer consider the orientation as an angle but as a vector on the d-dimensional sphere.

In order to justify the existence of the ridgelet transform, which we will recall afterwards, the mother rigelet should satisfy an admissibility property: $\int \frac{|\widehat{\psi}(\xi)|^2}{|\xi|^d}\mathrm{d}\xi < \infty$

Definition 2 (Two dimensional ridgelet transform). *Given an admissible ridgelet ψ and a function f, its ridgelet transform yields the following ridgelet coefficients:*

$$\mathrm{Ridge}_\psi f(s,t,\theta) := \langle f, \psi_{s,t,\theta}\rangle = \int \overline{\psi_{s,t,\theta}}(x,y)f(x,y)\mathrm{d}x\mathrm{d}y \tag{2.19}$$

and the function can be reconstructed with the following inversion formula:

$$f(x,y) = \int_0^{2\pi}\int_{-\infty}^{\infty}\int_0^{\infty} \mathrm{Ridge}_\psi f(s,t,\theta)\psi_{s,t,\theta}(x,y)\frac{\mathrm{d}s}{s^3}\mathrm{d}t\frac{\mathrm{d}\theta}{4\pi} \tag{2.20}$$

In this last expression, we would need to replace the measures in the integral with respect to the dimension in case of a higher dimensional signal.

It is also proven [Can98] that the ridgelet transform satisfies an equality similar to the Parseval equality. Moreover, it can be expressed by means of a Radon transform which allows efficient computations of the ridgelet transform thanks to the Fourier slice theorem.

A further extension consists in applying the ridgelet transform to the different sub-bands of the image obtained by a wavelet transform. This yields the curvelet transform, which is studied in details in [CD00, CD05a, CD05b]

Applications of such curvelets transform can be found for instance in the context of image denoising [SCD02] or image fusion [CKK04].

2.3.3 Denoising and compressing with wavelets

We detail here the basics of image denoising with wavelets. While this algorithm does not seem to fit perfectly in the framework of structures and irregularities in images, let us motivate this choice. Noise can be seen as unexpected deviation from a given pure signal. Whence the concept of irregularities. Moreover, as stated by Chang *et al.* [CYV00]

> The threshold acts as an oracle which distinguishes between the insignificant coefficients likely due to noise, and the significant coefficients consisting of important signal structures.

As it turns out the wavelet thresholding methods tend to remove noise while emphasizing structures. Note that this is how the JPEG2000 compression algorithm is developped [SCE01].

Without giving more details (which can be obtained for instance in [Mal89]) the thresholding method works as follows. Given a two-dimensional image f corrupted by some additive noise η (which is in general assumed to be Gaussian independent identically distributed), $f_n = f + \eta$, we compute a wavelet decomposition: $g_n = W_\psi f_n = W_\psi f + W_\psi \eta$ (note that this is a matrix and depends on the choice of the mother wavelet ψ). All coefficients but in the last low-resolution approximation are filtered so that small coefficients are set to 0. The filtered set of coefficients, $\widehat{g_n}$, is then transformed back in the space domain with the inverse wavelet transform. There are two main ways to filters those coefficients, for a certain threshold T:

- The soft thresholding method: $\widehat{g_n}(x) = \text{sign}(g_n(x)) \max\{|g_n(x)| - T, 0\}$
- The hard thresholding method: $\widehat{g_n}(x) = g_n(x) \cdot (|g_n(x)| > T)$

where the last product should be interpreted in a binary sense *i.e.* $(|g_n(x)| > T) = 1$ if the statement in the brackets is true or 0 elsewhere.

Note that other methods are possible but have come up later or based on the same idea. An important improvement is however to make an adaptive threshold instead of a global threshold for the whole image [CYV00].

In a first experiment, depicted on Fig. 2.6 we have applied a compression algorithm by increasing continuously the threshold.

Figure 2.6: Wavelet synthesis after hard (top row) and soft (bottom row) thresholdings with thresholds values (from left to right) of respectively 0.02, 0.06, 0.14, 0, 24 and 0.40.

As we can see on these images, within the first smallest thresholds (left images) the textural aspect of the bricks are suppressed. The structures start to get really distorted with higher thresholds. In an informal way we could say that the soft thresholding method tends to show some blurring effect while the hard thresholding one tends to create artifacts at the structures' borders (as it can be seen from the two most right images).

The next experiment is done by applying a threshold proportional to the noise variance and increase this noise variance. The results are depicted in Fig. 2.7. It comes out of these tests that up to a certain noise level, both soft and hard thresholding are able to keep track of the structures of the house. The brick wall textures are also kept with lower noise level. However whenever the noise is getting higher, its coefficients in the wavelet domain are getting bigger than the ones of the image underneath and therefore the denoising is somewhat weak. This is why an adaptive thresholding is needed for better applications.

Figure 2.7: Wavelet synthesis after hard (top row) and soft (bottom row) thresholdings in the presence of additive gaussian noise (images on Fig. 2.5). The threshold is proportional to the noise variance.

2.4 Phase-based image comparison

Another large research domain in the context of image and structure analysis can be found in the area of phase analysis. We first see how structures of an image can be found in the phase component of the Fourier transform, and how it can be used in diverse image processing and/or computer vision applications.

2.4.1 Phase as structural component of images

The whole phase-based image analysis theory has really started with the work of Oppenheim and Lim [OL81]. In there work they tried to motivate both heuristically and analytically the use of phase representation in imaging. We can for instance notice (an informal way) that the Fourier shift theorem is a motivation. Indeed, if the Fourier transform of a spatially translated image (or any general n-dimensional signals) can be computed by a shift of the phase in the Fourier domain. This process however keeps the spectral magnitude exactly the same. This tends to show that phases contain more structural or localisation information than the magnitude.

Another point pleading in favor of the importance of phase is when considering magnitude transformations before doing a Fourier synthesis. One can consider structures (edges and corners) as high frequency components of an image. Therefore, modifying the spectral magnitude in order to give more weight to higher frequency should emphasize the quality of edges. Unfortunately, it will also increase the presence of noise in the synthesised image.

From a more analytical point of view, we want to recall two theorems which can be used for a perfect reconstruction of signal only based on the phase.

Theorem 1 ([HLO80]). *A finite-length discrete signal which has a z-transform with no zeros on the unit circle nor in its conjugate reciprocal is uniquely specified to within a scale factor by the phase of its Fourier transform.*

Note that the z-transform can be understood as the discrete version of the Laplace transform.

Definition 3. *Given a discrete time signal* $\{f_k\}_k$, *its z-transform is defined as*

$$F(z) = \mathcal{Z}\{f_k\}(z) = \sum_{k=-\infty}^{\infty} f_k z^{-k} \qquad (2.21)$$

In other words, the previous theorem gives a condition under which the phase of the Fourier spectrum is sufficient for a perfect reconstruction. This theorem can also be generalised to higher dimensional signals.

We have another similar theorem for perfect reconstruction.

Theorem 2 ([Hay82]). *A discrete signal of length $N < \infty$ which has a z-transform with no zeros on the unit circle or in conjugate reciprocal is uniquely specified to within a scale factor by $N-1$ samples in the interval $0 < \omega < \pi$.*

We illustrate in the next section this importance of phase in imaging, by playing with different spectrum magnitudes.

2.4.2 Phase-based applications in image processing

Here we want to give some basic examples of how we can use the phase for image processing applications. These examples will be as simple as posible in order to emphasize the importance of phase. It is not intended as a state-of-the-art research.

Image swapping   We prove in this paragraph, that most of the structural information is contained in the phase and not in the amplitude of the Fourier spectrum. As an illustrative example, we consider 4 common images for computer vision scientist (in Fig. 1.2(a), 1.2(b), 1.2(c), and 1.2(d), which we denote by Lena, Barbara, Mandril and Mansion images respectively). The second column of Fig. 2.8 displays the logarithm of the amplitude of the Fourier transform while the third one is its phase. In the last column, we have mixed the phase information of an image with the amplitude information of another. On the first row (Fig. 2.8(d)), the phase of Barbara is synthetised with the amplitude of Lena. In the last three rows, we use the phase of Lena with the amplitude of Barbara (Fig. 2.8(h)), the Mandril (Fig. 2.8(l)) and the House (Fig. 2.8(p)) images.

It is interesting to notice that even if the images created by mixing information from two different references are really noisy, we can still recognise the image from which the phase information is taken (*i.e.* Barbara on the first row and Lena and the three following images).

Amplitude corrections   As an illustrative example, we see how we can slightly modify an image by working on the amplitude of the Fourier domain but keeping the phase unchanged. We will illustrate this on two use-cases. First, in the first row of Fig. 2.9, the Lena image is corrupted by a dark blob-like background centered at the middle of the image. In a second experiment, the image is being corrupted by some Gaussian noise.

The first column of the Figure shows the original image with the appropriate distortion. Then from left to right, the correction are done on the amplitude by 1) cutting off all the frequency above a certain threshold (in the illustrated experiments, we have cut off the higher half of the frequency domain) and leaving all the other unchanged, 2) multiplying the amplitudes by some Gaussian weights

2.4. PHASE-BASED IMAGE COMPARISON

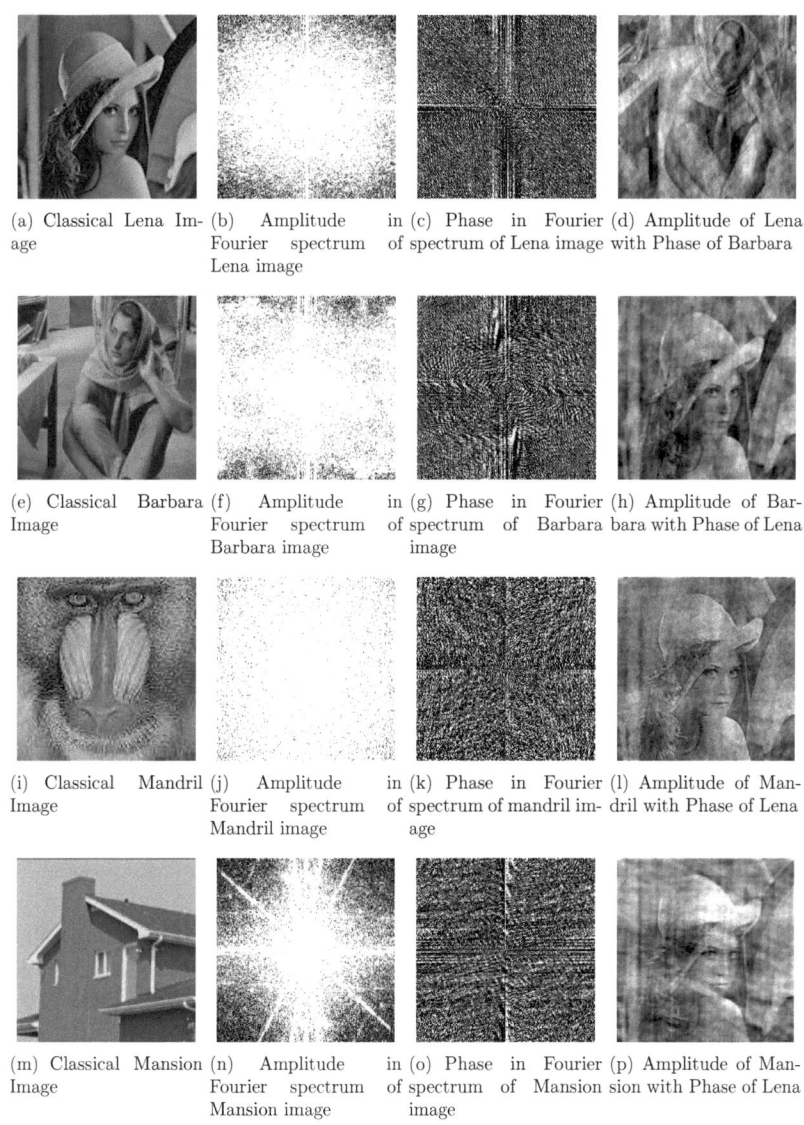

(a) Classical Lena Image (b) Amplitude in Fourier spectrum Lena image (c) Phase in Fourier of spectrum of Lena image (d) Amplitude of Lena with Phase of Barbara

(e) Classical Barbara Image (f) Amplitude in Fourier spectrum Barbara image (g) Phase in Fourier of spectrum of Barbara image (h) Amplitude of Barbara with Phase of Lena

(i) Classical Mandril Image (j) Amplitude in Fourier spectrum Mandril image (k) Phase in Fourier of spectrum of mandril image (l) Amplitude of Mandril with Phase of Lena

(m) Classical Mansion Image (n) Amplitude in Fourier spectrum Mansion image (o) Phase in Fourier of spectrum of Mansion image (p) Amplitude of Mansion with Phase of Lena

Figure 2.8: First easy tests to show the importance of using the phase in image processing. First column shows some examples of natural image scenes. Second and third columns correspond resp. to their amplitude and phase in Fourier domain. Last columns is produced by inverse Fourier transform a combination of phase and amplitude coming from two different images.

Figure 2.9: Effect of amplitude modification in the Fourier domain of an image. In the first row, the Lena image is corrupted with a blob-like background. In the second one, Gaussian noise is added to the original image.

centered at the 0 frequency (*i.e.* the highest frequencies get less weight), 3) multiplying by a linear weight with value 1 at the centre and 0 at the border of the frequency domain and 4) setting all amplitudes to 1.

As one would expect from basic Fourier theory the edge and Gaussian low pass filters have very similar effects. The triangular filter seems to improve the contrast which leads to a darker but smaller blob on the eye. The unit amplitude image keeps every, and only, fine structures of the image but is therefore very sensitive to the presence of noise.

2.5 Mathematical morphology

The mathematical morphology acts as an important example of structural image processing. It is based on different algebraic operations (such as opening, closing, dilation and erosion) and is controlled by what is called a *structuring element*. This structuring element entails the features one wish to detect in a signal. It finds some application for instance in texture analysis and defect detection [HM96], satellite image segmentation [PB01], SAR image processing [OHCR96] or in medical image analysis [MFB+95].

This section is only intended as a short overview of what can be done and what is being done using mathematical morphology. For more details it is suggested to read [HSZ87, HR90, RH91]. We first describe the operations based on mathematical morphology for binary images and introduce its extension to sampled functions.

2.5.1 Mathematical morphology on binary images

As we will detail better in a later chapter (see Chap. 5), we can consider a (discrete) binary image as a set A in $\mathbb{Z}^2$. All operations in mathematical morphology require a structuring element (which is generally denoted by the letter B). For instance, we can consider a cross defined on a 3×3 grid. An example of image and structuring element are given in Fig. 2.10

Four main operations (illustrated on Fig. 2.11 for the examples 2.10) are often used in mathematical

2.5. MATHEMATICAL MORPHOLOGY

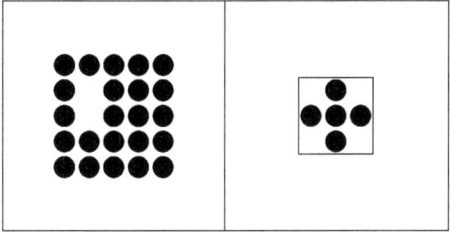

Figure 2.10: Example of binary image (left) and structuring element (right).

morphology:

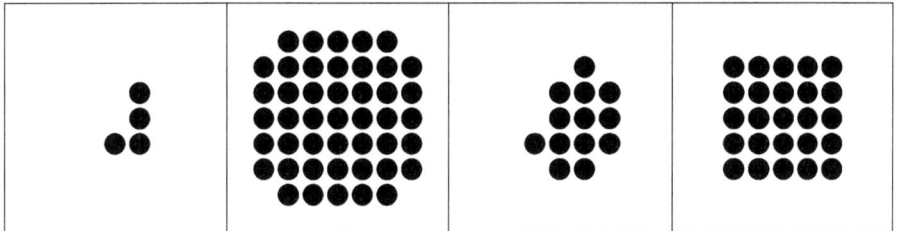

Figure 2.11: Examples of binary morphological operations. From left to right: Erosion, Dilation, Opening, Closing

Erosion It works by sequentially translating the structuring elements on all pixels of the binary image. Whenever the structuring element is fully contained in the image, the current central pixel is kept in the resulting image. More formally we have

$$A \ominus B := \{p \in A : B_p \subset A\} \quad (2.22)$$

where $B_p = \{b + p : b \in B\}$ represents the translated of B centered at pixel p.

Dilation It is somehow the opposite to the previous operation where the binary image gets extended according to the structuring element. It means that for each pixel from the image, we overlap the structuring element centered on that pixel and mark all the pixels beneath as part of the resulting image. Formally speaking:

$$A \oplus B := \cup_{a \in A} B_a \quad (2.23)$$

Opening This operation is done by successively applying an erosion and then a dilation.

$$A \circ B := (A \ominus B) \oplus B \quad (2.24)$$

In a certain way we can say that the opening keeps the pixels which belong to a translation of the structuring element centered on a pixel of the image and which are fully contained in the image. It tends in another way to keep the whole structure which is completely inside an image, depending on a given structuring element.

Closing The morphological closure is computed by applying first the dilation operator and then the erosion one

$$A \bullet B := (A \oplus B) \ominus B \qquad (2.25)$$

The closing operation tends to cover the holes which are contained in an image but which are of a size in the order of the structuring element.

As we can see from Fig. 2.11 the erosion operator emphasizes the parts which are "densely" contained in the image, the dilation one extends the object according to the structuring element, the opening operator gets rid of the tiny parts of the image, while the closing one gets rid of the tiny holes in the images (it literally closes the images).

2.5.2 Extension to sampled functions

An extension to sampled image is done as a non linear filtering as follows:

Definition 4 (Morphological operations of gray scaled images). *If f and b are functions defined (for an image) on a subset $D \subset \mathbb{R}^2$ with values in $\mathbb{R}$ we can define*

$$(f \oplus b)(x) := \sup_{y \in D} \left(f(y) + b(x - y) \right) \qquad (2.26)$$

$$(f \ominus b)(x) := \inf_{y \in D} \left(f(y) - b(y - x) \right) \qquad (2.27)$$

$$(f \circ b)(x) := \left((f \ominus b) \oplus b \right)(x) \qquad (2.28)$$

$$(f \bullet b)(x) := \left((f \oplus b) \ominus b \right)(x) \qquad (2.29)$$

Fig. 2.12 illustrates these operations on the House image with different structuring elements (a small disk, two rectangles of the size of the bricks or of the windows).

As we can see the structuring element plays an important role in the outcome of the morphological operations. The operations tend to react to features in the size order and shape of the structuring element. This becomes evident in the last row where only large rectangles appear in the outputs due to the structuring element having the shape and size of the windows of the house.

Moreover one can defined gradient like operators (depicted on Fig. 2.13 for the house image)

Definition 5 (Morphological gradient and Laplacian). *As we would for normal image operations we can define a gradient as*

$$\operatorname{Grad}_b(f)(x) := (f \oplus b)(x) - (f \ominus b)(x) \qquad (2.30)$$

and a Laplacian

$$\operatorname{Lap}_b(f)(x) := (f \oplus b)(x) + (f \ominus b)(x) - 2f(x) \qquad (2.31)$$

Other more complicated operations can be defined but are not given any importance here.

As we can notice the brick structuring element gives much lower responses to horizontal edges compared to the disk element (which can be seen as an isotropic edge detector). This effect can also be seen in the Laplacian images (second and last one in the figure).

2.5. MATHEMATICAL MORPHOLOGY

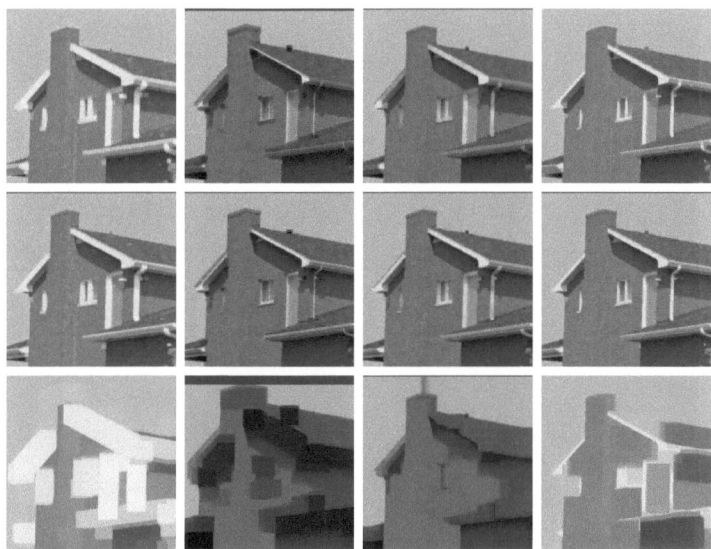

Figure 2.12: Basic morphological operations on gray-valued images. First row makes use of a small disk-shaped structuring elements while the two following ones use rectangle structuring elements with in the order of the size of a brick or of a window respectively. Columns left to right show the dilation, erosion, opening and closing operations.

Figure 2.13: Morphological gradient and Laplacian using a disk shaped structuring element (first two columns) and a small rectangle structuring element (last two columns).

Part II

Analysis of irregularities with the discrepancy norm

Chapter 3

Theoretical Aspects of the Discrepancy Norm

This chapter is dedicated to the theoretical study of the discrepancy norm. As we see the discrepancy norm arises from the early years of the 20th century in the context of divergence of sequences. Since then it has been applied to different areas in mathematics and in particular in stochastic analysis and was recently rediscovered as an interesting similarity measure for images.

We start in Section 3.1 by recalling what the historical foundations of the discrepancy theory are. Then we give in Section 3.2 its more recent formulation for general μ measurable functions. Some characteristic properties such as a Lipschitz like continuity and how this property is quite remarkable regarding other known (dis-)similarity measures are introduced

3.1 A century of mathematical research about the divergence of sequences

3.1.1 Introduction

The discrepancy theory appeared with the pioneer work of Weyl [Wey16]. He stated the problem as follows: given a sequence of infinitely many real numbers $\{x_k\}_k$, if we roll them up on a unit length circle, would they densely cover this circle? Formally speaking, we consider the equivalence relation "equals modulo 1" and say that $x = y$ mod. $1 \Leftrightarrow \exists m \in \mathbb{Z} : x = y + m$. Therefore we can constrain our family of numbers to be in $[0, 1)$. In this case, $\forall k \in \mathbb{N}, x_k$ corresponds to the coordinate of the rolled up point on the circle.

If we consider the set of arcs Ξ and denote by $\ell(\Theta)$ the length of an arc $\Theta \in \Xi$ on this circle, the set of rolled up numbers is said to densely cover the circle if

$$\forall \Theta \in \Xi, \lim_{N \to \infty} \frac{|\{x_k\}_{k=1}^N \cap \Theta|}{N} = \ell(\Theta) \qquad (3.1)$$

Definition 6 (Discrepancy of a sequence). *Let $x = \{x_k\}_{k \in \mathbb{N}}$ be a sequence of real number in the unit* interval and $N \in \mathbb{N}$ an integer. The discrepancy of order N of x reads

$$D_N(x) = \sup_{0 \leq a \leq b \leq 1} \left| \frac{|\{x_k\}_{k=1}^N \cap [a,b]|}{N} - (b-a) \right| \quad (3.2)$$

Property 1 (Equidistribution of a sequence). *The sequence $\{x_k\}_k$ is said to be equidistributed if* $\lim_{N \to \infty} D_N(x) = 0$

Equidistribution is equivalent to the dense covering of the unit-length circle of points whose coordinates on the circle are x_k

3.1.2 Domains of applications

Since the original work of Weyl [Wey16], discrepancy theory's interest has increased. Many applications and development have occurred. We here only give some examples of applications with some references for the curious reader. For a rather complete and interesting review of discrepancy methods, we suggest to read the book of Chazelle [Cha01].

A first development based on Hermann Weyl's was the work of Koksma [Kok35]. In this work the author was interested in studying the pseudorandomness of the exponential function and how its distribution modulo 1 would densely cover the unit circle. Such irregularities or uniformity have also been studied in [KN05, BC09].

Discrepancy methods appear in the estimation of Monte Carlo methods. In this area, authors [Zar00, Nie92] want to assess the quality of the convergence of Monte-Carlo methods seen as an empirical representation of a continuous distribution. This concept perfectly fits to our definition of discrepancy, Eq. 3.2.

Of particular interest to us are applications in the domain of image processing. The discrepancy theory already appeared for image recognition [NW87], pixel classification [BBK96] or for the analysis of digital halftoning of images [SCT02].

3.2 The modern approach to n-dimensional signals

We analyse now the approach due to Moser [Mos11] for images and volumetric data which leads to the definition of the discrepancy norm we are using in this work. We introduce in particular its interest in the context of discrepancy-based autocorrelation functions, as this is what is needed in for instance image or time series alignment.

3.2.1 Similarity and misalignment

The concept of discrepancy norm for image analysis as we use it in the rest of this manuscript was mathematically described in [Mos11] for the purpose of measuring the extent of misalignment between signals. The work on the discrepancy norm appears to be interesting to the author based on an axiomatic description of similarity measures:

[C1] a vanishing distance entails a vanishing extent of misalignment and vice versa (which we refer to as *positive-definiteness*)

3.2. THE MODERN APPROACH TO N-DIMENSIONAL SIGNALS

[C2] the distance measure behaves continuously at least with respect to arbitrary small misalignments (*continuity*)

[C3] an increasing extent of misalignment implies an increasing distance measure and vice versa (*monotonicity*)

Based only on these three criteria, one can see that the naturally appearing similarities practically used fail. Criteria [C3] is actually very constraining. We see in a later section (3.3.2) that we can actually derive a construction principle for similarity measures that allows one to generalize many well-known similarities and that none of these similarities will fulfil criteria [C3].

However, before giving more technical details, we introduce some definition which will be useful along this chapter.

Definition 7 (Autocorrelation function). *Given a multidimensional signal f with domain in $\mathbb{R}^n$ and values in $\mathbb{R}^q$, i.e. $f : \mathbb{R}^n \to \mathbb{R}^q$, and an error metric d the autocorrelation function of signal f is defined as:*

$$\Delta_d[f](t) = d(f, \tau_t f) \tag{3.3}$$

In a similar way, one can define the correlation function between two signals f and g as

$$\Delta_d[f,g](t) = d(f, \tau_t g) \tag{3.4}$$

note that we can independently write $d(f, \tau_t g)$ or $d(\tau_{t'} f, g)$ just by reversing the shift axis: $t' \leftarrow -t$

Now using this notation, we can model all three criteria [C1]-[C3] as follows:

[C1] $\Delta_d[f](t) = 0 \Leftrightarrow t = 0$. It means that a perfect score could and should be achieved only when two exact same signals are being compared.

[C2] $\Delta_d[f](t) \xrightarrow[t \to 0]{} 0$. This means that a similarity measure should smoothly approach a perfect match when the misalignment tends to 0; no jumps at that point are allowed.

[C3] $\forall t \in \mathbb{R}^n, \forall \lambda_1, \lambda_2 \in \mathbb{R}, 0 \leq \lambda_1, \leq \lambda_2 \Rightarrow \Delta_d[f](\lambda_1 t) \leq \Delta_d[f](\lambda_2 t)$, which means that increasing the misalignment in a direction should increase the autocorrelation function.

Some examples These constraints seem general. However, it looks like only a few similarity measures will fulfil them. We only give here some examples with a use case causing the similarity measure to fail in the monotonicity property of the autocorrelation function. More details are given later in Sec. 3.3.2 regarding construction principles of similarity measures.

One of the most used similarity measure (for instance known as the Gaussian kernel in *Support Vector Machines* (SVM) [CST00], in the machine learning community) is computed as the inverse of the exponential of the Euclidean distance:

$$s_G(f, g) = e^{-\frac{\|f-g\|_2^2}{\sigma_G}} \tag{3.5}$$

This similarity is widely used for many reasons:

- Its values range from 0 (in a limit case) for completely dissimilar patterns f and g (i.e. $\|f-g\|_2^2 \to \infty$) and 1 only for perfect match (i.e. $f \equiv g$).

- It is usable 'as is' in the SVM framework due to its positive definiteness (it is said to be a Mercer kernel, or to fulfil Mercer's condition):

$$\forall g \in L^2, \iint s_G(x,y) g(x) g(y) \mathrm{d}x \mathrm{d}y \geq 0 \qquad (3.6)$$

- The "closeness" can be tuned optimally to a given application through the radius σ_G of the kernel: $\sigma_G \to 0$ tends to give strong discrimination while $\sigma_G \to \infty$ tends to give higher similarities, what we can also interpret as more invariance during the classification process.

- Its computation can be done very efficiently due to the scalar product and to its separability in each dimension.

However, if one considers the following function $f^* := \mathbf{1}_{[1,2)} + \mathbf{1}_{[3,4)}$ and $\Delta_G[f](t) := s_G(f - \tau_t f)$ with unit radius σ_G, we have the similarity values given in Tab. 3.1.

Table 3.1: Some counter examples for the monotonicity of the gaussian kernel autocorrelation function

t	0	1/2	1	3/2	7/4	2	∞
$\Delta_{L^2}[f^*](t)$	0	2	4	3	3.5	3	4
$\Delta_G[f^*](t)$	1	0.1353	0.0183	0.0498	0.0302	0.0498	0.0183

This Table 3.1 shows another interesting dissimilarity: the L^2 norm. It shows the similar strongly non-monotonic behaviour (as expected, due to the monotonicity of the exp function). This example can also be applied to other Minkowski metrics to show the non-monotonicity of their autocorrelation functions.

Another often used similarity measure in the context of image registration is the well-known mutual information [WIVA+96, VWI97]. Once again, it can be shown by an easy example, that aligning two patterns with this similarity measure is not a trivial task (mainly due to the fact that it is based on probability, and therefore does not account any spatial information), and one wants to be careful during the optimisation process (done in those papers using a stochastic gradient descent [BB08]).

3.2.2 The one dimensional case

3.2.2.1 Definition

If we come back to the definition of the discrepancy measure of a sequence $x = \{x_k\}_k$:

$$D_N(x) = \sup_{0 \leq a \leq b \leq 1} \left| \frac{|\{x_k\}_{k=1}^N \cap [a,b]|}{N} - (b-a) \right|$$

This problem can also be seen as the divergence between two measures. The first one $N(a,b) = |\{k, 1 \leq k \leq N : u_k \in [a,b]\}|/N$ is actually the empirical distribution of a measure μ given by the samples x_k while $\nu(a,b) = b - a = \int_a^b 1$ is the uniform measure on $[a,b]$. With this new insight, we can write the discrepancy measure as

$$D(\mu, \nu) = \sup_{0 \leq a \leq b \leq 1} |\mu(a,b) - \nu(a,b)| \qquad (3.7)$$

3.2. THE MODERN APPROACH TO N-DIMENSIONAL SIGNALS

Now as seen in [Sch50], a signal (which we consider as a compactly supported function or eventually defined on the torus for periodic cases) can be seen as a measure:

Property 2 (Measure and generalized functions - Schwartz). *A measure is a concept that generalizes the concept of function and there is a univocal correspondence between the class of measure absolutely continuous and the set of compactly supported functions:*

$$\forall \phi \in C_c, \ \mu(\phi) := \int_{\mathbb{R}^n} f\phi \tag{3.8}$$

This result can be shown by having a look at the density of a given measure μ on one side and at the integral of a function f over a set on the other. More details can be found in [Sch50]. In this case, we call μ a *measure* and f a *function* or the *density* of measure μ.

Finally, if we consider a signal as being a compactly supported function, we can define the discrepancy between any two signals f and g as:

$$D(f,g) = \sup_{a,b \in \mathbb{R}} |\int_a^b f - \int_a^b g| \tag{3.9}$$

This equation measures the maximum deviation of a signal from another one. If we consider the case where g is the constant 0 distribution, we can get our definition for the discrepancy of a signal as:

Definition 8 (Discrepancy of a signal). *We define the discrepancy of any compactly supported signal f on a σ finite algebra with measure μ as*

$$D(f) = \sup_{a,b \in \mathbb{R}} |\int_a^b f \mathrm{d}\mu| \tag{3.10}$$

or, for a discrete finite signal

$$d(f) = \max_{a,b \in \mathbb{Z}} |\sum_{l=a}^{b} f_l|$$

And we prove the following result:

Property 3 (Norm property). *The formula for the discrepancy of a signal defined above fulfils the axioms of a norm.*

Proof. To show the axioms of a norm of the previous formula let us rewrite it differently. To this task we write $F(a,b) = \int_a^b f \mathrm{d}\mu$. It is clear that we have $D(f) = \|F\|_\infty$. Remark however that this ∞ norm is taken on $\mathbb{R}^2$. It is now obvious that $D(\cdot)$ is norm on $L^1(\mathbb{R}, \mu) \cap L^\infty(\mathbb{R}, \mu)$ (mainly due to the linearity of the integral). □

From now on, we only talk about the discrepancy norm and use the notation $\|\cdot\|_D$ whenever needed. To our knowledge the first use of the discrepancy measure in this form is in [NW87] where the authors address the problem of pattern recognition through distances in metric spaces. However, it lacked all the theory until the mid-90s and the work of Bauer et al. [BBK96] who first proved the norm axioms and used the discrepancy norm in the discrete case for edge detection in image processing. Later, Moser [Mos11] proved the norm axioms for the general continuous measurable case.

However it is not straightforward how to use this discrepancy norm for higher dimensional objects. Indeed both works [NW87, BBK96] were working on images (or parts of an image) but decided to handle it as a vector by unrolling the matrix (columnwise for Neunzert et al. and as a circle or spiral-like for Bauer et al.). This is clearly not the best we can do as, according to the definition of the discrepancy norm, the ordering of the elements is important. For instance, if we consider a vector $f_1 = (1, 1, -1)$ and another one $f_2 = (1, -1, 1)$, we have that $\|f_1\|_D = 2 \neq 1 = \|f_2\|_D$. This ordering has its importance as it shows how the different intensities are organized in the space. This means that the discrepancy norm tends to emphasise the biggest structure of the objects we are comparing. For all these reasons, an extension to higher dimensional cases was mandatory for practical examples and are detailed in Section. 3.2.3.

3.2.2.2 Computational aspects

Referring to Eq. 3.10 and its discrete counterpart, we see that it is at the moment computationally greedy, as it has a complexity in $\mathcal{O}(N^2)$, with N being the number of data samples. Fortunately, it can be implemented in a smarter way.

Property 4 (Fast computation of the discrepancy norm). *The discrepancy norm of a signal f can be computed in linear time using the following equivalent formulation:*

$$\|f\|_D = \sup_{a \in \mathbb{R}} \int_{-\infty}^{a} f \mathrm{d}\mu - \inf_{a \in \mathbb{R}} \int_{-\infty}^{a} f \mathrm{d}\mu \qquad (3.11)$$

Or for discrete vectors:

$$\|f\|_D = \max_{k \in \{1, \cdots, N\}} \left\{ 0, \sum_{l=1}^{k} f_l \right\} - \min_{k \in \{1, \cdots, N\}} \left\{ 0, \sum_{l=1}^{k} f_l \right\} \qquad (3.12)$$

Remark that the last formula seems strange and somewhat arbitrary, but it should be understood that the sums are actually indexed on $\mathbb{Z}$. But the signals being summable implies that their coefficients vanish at infinity whence the sequence starting with a 0.

Proof. The proof of this proposition is rather straightforward.

$$\sup_{a,b \in \mathbb{R}} \left| \int_a^b f \mathrm{d}\mu \right| = \sup_{a,b \in \mathbb{R}} \left| \int_{-\infty}^b f \mathrm{d}\mu + \int_{-\infty}^a -f \mathrm{d}\mu \right|$$
$$= \sup_{b \in \mathbb{R}} \int_{-\infty}^b f \mathrm{d}\mu + \sup_{a \in \mathbb{R}} \int_{-\infty}^a -f \mathrm{d}\mu = \sup_{b \in \mathbb{R}} \int_{-\infty}^b f \mathrm{d}\mu - \inf_{a \in \mathbb{R}} \int_{-\infty}^a f \mathrm{d}\mu$$

The second equality is due to the fact that $f \in L^1$ so that $\sup_{b \in \mathbb{R}} \int_{-\infty}^b f \mathrm{d}\mu \geq 0$ and $\sup_{a \in \mathbb{R}} \int_{-\infty}^a -f \mathrm{d}\mu \geq 0$ and that the sup over a and b is taken completely independently from each other. □

With this formulation and the use of integral images [Cro84, VJ01], the discrepancy norm can be computed in linear time by computing the cumulative integrals and memorizing both highest and lowest values.

3.2.3 Higher dimensional generalization

Unfortunately, the generalization to higher dimensional spaces is not straightforward. While in the one dimensional case, the suprimum is taken over the set of all possible intervals, some choice has to be done regarding the set over which to maximize the discrepancy. Some variants are proposed in the following sections.

3.2.3.1 On connected components

A natural generalization of intervals to higher dimensional spaces are the compact subsets of $\mathbb{R}^n$. In this section we denote by $\mathcal{C}$ the set of all compact subsets of $\mathbb{R}^n$.

Definition 9 (Discrepancy norm on compact sets). *The discrepancy norm can be extended to higher dimensional spaces by maximizing the discrepancy over compact sets. Let f be a n-dimensional signal ($n \geq 1$), we define the following*

$$\|f\|_{\mathcal{C}} := \sup_{c \in \mathcal{C}} \left| \int_c f \mathrm{d}\mu \right| \tag{3.13}$$

While this definition seems robust as it allows to consider many possible subsets of the input space, it is unfortunately untractable on concrete applications. Fig 3.1 shows some of the many possible admissible sets appearing as candidate in the supremum

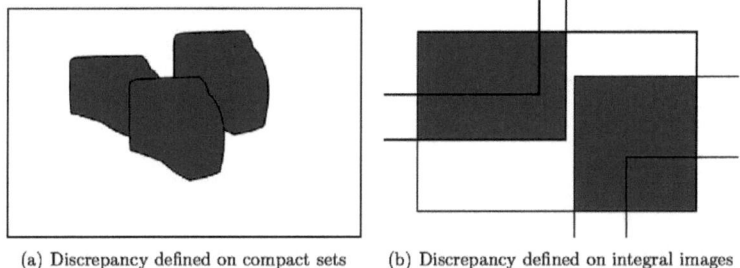

(a) Discrepancy defined on compact sets   (b) Discrepancy defined on integral images

Figure 3.1: Examples of admissible sets for the definition of the discrepancy norm.

3.2.3.2 Box

In this case, we consider only sets that are Cartesian products of (eventually infinite) intervals in each dimension $\mathcal{B} = \{\prod_{i=1}^n (a_i, b_i), \forall i \in \{1, \cdots, n\}, a_i \in \overline{\mathbb{R}}, b_i \in \overline{\mathbb{R}}, \text{ and } a_i \leq b_i\}$

Definition 10 (Discrepancy on Cartesian product of intervals). *f being similar as in the previous definition, we can define the discrepancy norm on Cartesian products of intervals:*

$$\|f\|_{\mathcal{B}} := \sup_{B \in \mathcal{B}} \left| \int_B f \mathrm{d}\mu \right| \tag{3.14}$$

This offers yet a good approximation (compact sets can be approximated by such boxes) of the previous one and much lower number of candidates, it is still not satisfying in terms of computational complexity.

3.2.3.3 Infinite box

Here we consider a subset of the previous set of boxes where at least one end of each interval is infinite. Formally, we define by $\widetilde{\mathcal{B}}$ the set of all Cartesian product of intervals of the form $(-\infty, b)$ or (a, ∞)

Definition 11 (Discrepancy on Cartesian products of infinite boxes). *f being as above, we have another potential definition of the discrepancy norm*

$$\|f\|_{\widetilde{\mathcal{B}}} := \sup_{B \in \widetilde{\mathcal{B}}} \left| \int_B f \, \mathrm{d}\mu \right| \tag{3.15}$$

3.2.3.4 As a difference of suprimum and infimum

In this part we try, instead of generalizing the earlier definition we try to generalize Eq. 3.11 by computing the difference between the suprimum and the infimum.

And this case however, and due to the multidimensionality of the input space, many directions of integration are possible (2^n for a n-dimensional space). Let us first introduce some notations to make the rest easier. Let $\iota \in \{-1, 1\}^n$ denote a direction for integration and for $s \in \mathbb{R}^n$, $[s[:= (-\infty, s_1) \times \cdots \times (-\infty, s_n)$. For a given ι, we denote by $\iota \cdot [s[$, the set composed of a product of n integrals: $\iota \cdot [s[:= (-\iota_1 \infty, \iota_1 s_1) \times \cdots \times (-\iota_n \infty, \iota_n s_n)$. In terms of integrals it means

$$\|f\|_I^{(n)} := \max_{\iota \in \{-1,1\}^n} \left\{ \sup_{s \in \mathbb{R}^n} \int_{\iota \cdot [s[} f \, \mathrm{d}\mu - \inf_{s \in \mathbb{R}^n} \int_{\iota \cdot [s[} f \, \mathrm{d}\mu \right\} \tag{3.16}$$

where each integrals might have to be computed backwards.

Due to their definitions, the two last formulae can moreover be implemented with the integral image concept.

3.2.3.5 Equivalence of the autocorrelation functions

Here we prove that there is no need to make special considerations about which definition to take when considering the problem of misalignment of signals.

Theorem 3 (Equivalence of the misalignment functions, Moser). *For any dimension $n \geq 1$ and positive measurable functions $f \in L^1(\mathbb{R}^n, \mu)$ it holds:*

$$\|f - \tau_t f\|_{\mathcal{B}} = \|f - \tau_t f\|_{\widetilde{\mathcal{B}}} = \sup_{\iota \in \{-1,1\}^n} \sup_{s \in \mathbb{R}^n} \int_{\iota \cdot [s[} f - \tau_t f \, \mathrm{d}\mu \tag{3.17}$$

The proof of this result can be found in [Mos11] and is based on extending the boxes from one of the corners to the infinity in the opposite direction. Then dealing with translations of such extended boxes concludes the result.

3.3 Characteristic properties

In this section we give some remarkable properties of the discrepancy norm. We first see that the discrepancy autocorrelation function fulfils a Lipschitz continuity condition: the higher the misalignment

3.3. CHARACTERISTIC PROPERTIES

in a direction, the higher the discrepancy, and this discrepancy is uniformly bounded by the infinity norm. Then we see that under some rather general assumptions, classical similarity measures cannot fulfil this property. We also study the case of probabilistic f-divergences.

3.3.1 A Lipschitz property

The two theorems here are given without proofs, which can be found in [Mos11]

Theorem 4 (Monotonicity of the misalignment function). *Let $0 \leq \lambda_1 \leq \lambda_2$ be two real numbers, and $f \geq 0$ be a non-negative function, it holds*

$$\forall t \in \mathbb{R}^n \Delta_D[f](\lambda_1 t) \leq \Delta_D[f](\lambda_2 t) \tag{3.18}$$

Theorem 5 (Lipschitz property of the misalignment function). *The discrepancy correlation fulfils the following Lipschitz like inequality. Let f be a non negative measurable function defined on $\mathbb{R}^n$ and let $\delta_\mu[f](t) = \sup_{B \in \mathcal{B}} \max\{\mu(B \backslash \tau_t(B)), \mu(\tau_t(B) \backslash B)\}$. It holds*

$$\forall t \in \mathbb{R}^n, \Delta_D[f](t) \leq \delta_\mu[f](t) \|f\|_\infty \tag{3.19}$$

3.3.2 Non-monotonicity of misalignment functions

In this section we introduce a construction principle for similarity measures that is both general and weak when analyzing misaligned functions.

3.3.2.1 A construction principle

We start with some general construction routines for generating admissible similarity measures. As a first example, let us consider the scalar product of two vectors:

$$\langle f, g \rangle = \sum_i f_i g_i \tag{3.20}$$

This can be used in the context of misalignment measure (*i.e.* when g is obtained after a translation of f) where the response has its highest peak when the two patterns are perfectly aligned. It is actually constructed by a combination of a coordinate-wise operation (a product in this case) and an aggregation function (summing up all contributions). This is the start of our construction. We consider similarity measures created by combining a coordinate-wise operation with an aggregation function:

$$S_{\mathcal{A}, \mathcal{X}}(f, g) := \mathcal{A}\left(\mathcal{X}\left(f(x), g(x)\right)\right) \tag{3.21}$$

where $\mathcal{X}$ and $\mathcal{A}$ represent the coordinate-wise operation and the aggregation function respectively. The two functions f and g are taken in some admissible space X (for instance in the case of the scalar product, $\mathbb{R}^n$ or ℓ^2 would be perfect).

Eventually, to be more exhaustive, a monotonic scaling function s should be added as a last step in the construction principle to allow more general formulations:

$$S_{\mathcal{A}, \mathcal{X}, s}(f, g) := s\left(\mathcal{A}\left(\mathcal{X}\left(f(x), g(x)\right)\right)\right) \tag{3.22}$$

First examples We give here some small general examples to motivate the choice of such constructions.

The Gaussian kernel which we have already introduced can be constructed by such principles using

- $\mathcal{X}(f(x), g(x)) = f(x) - g(x)$, as a component-wise operation,
- $\mathcal{A}(h) = \sum_i h_i^2$, as an aggregation,
- $s(v) = exp(-\frac{v}{\sigma_G})$ as a scaling function,

and as we have seen earlier, this kernel does not fulfil the monotonicity property for its autocorrelation function.

Note furthermore that using the square root as a scaling function gives the classic Euclidean norm. This dissimilarity does not fulfil the monotonicity property either.

3.3.2.2 A non-monotonicity property

Theorem 6 (Non monotonicity of autocorrelations [MSB11]). *Given a coordinate wise operation, an aggregation and a scaling function according to the construction above, the autocorrelation function based on such (dis-)similarity measure does not fulfil the monotonicity property under some conditions. More formally, if we have*

$$\Delta_{\mathcal{X},\mathcal{A},s}[f,g](t) := s\left(\mathcal{A}\left(\mathcal{X}(f(x), g(x-t))\right)\right) \quad (3.23)$$

where f and g are functions defined on $\mathbb{Z}$ and $\mathcal{X}$ is the coordinate-wise operation satisfying

(C1) *$\mathcal{X}$ is commutative,*

(C2) *$\mathcal{X}(1,0) \neq \mathcal{X}(1,1)$,*

(C3) *$\mathcal{X}(0,0) = \min\{\mathcal{X}(1,0), \mathcal{X}(1,1)\}$,*

and the aggregation function $\mathcal{A}$ is

(A1) *commutative and*

(A2) *strictly monotonically increasing, respectively decreasing, in each component,*

and the scaling function s is strictly monotonically increasing, respectively decreasing, then

$$\exists f : \mathbb{Z} \to \mathbb{R}, t_1, t_2, t_3 \in \mathbb{Z} : 0 \leq t_1 < t_2 < t_3 \Rightarrow \quad (3.24)$$

$$\begin{cases} \Delta_{\mathcal{X},\mathcal{A},s}[f,f](t_1) \leq \Delta_{\mathcal{X},\mathcal{A},s}[f,f](t_2) \geq \Delta_{\mathcal{X},\mathcal{A},s}[f,f](t_3) \\ \Delta_{\mathcal{X},\mathcal{A},s}[f,f](t_1) \geq \Delta_{\mathcal{X},\mathcal{A},s}[f,f](t_2) \leq \Delta_{\mathcal{X},\mathcal{A},s}[f,f](t_3) \end{cases} \quad (3.25)$$

and therefore is the monotonicity of the autocorrelation function not fulfilled.

Proof. To simplify, we consider the case where the aggregation function $\mathcal{A}$ is strictly increasing. The other case can then be done similarly.

3.3. CHARACTERISTIC PROPERTIES

Let's consider the function f defined on $\mathbb{Z}$ as

$$f(x) = \mathbf{1}_{\{0\}}(x) + \mathbf{1}_{\{2\}}(x) \tag{3.26}$$

and, for simplicity reasons again, let $\mathcal{X}_{00} = \mathcal{X}(0,0)$, $\mathcal{X}_{01} = \mathcal{X}(0,1)$, $\mathcal{X}_{10} = \mathcal{X}(1,0)$, $(= \mathcal{X}_{01}$ due to the commutativity) and $\mathcal{X}_{11} = \mathcal{X}(1,1)$
and consider, for $t_1 = 0$, $t_2 = 1$ and $t_3 = 2$

$$\Delta_0 = \Delta_{\mathcal{X},\mathcal{A},s}[f,f](t_1), \tag{3.27}$$
$$\Delta_1 = \Delta_{\mathcal{X},\mathcal{A},s}[f,f](t_2), \tag{3.28}$$
$$\Delta_2 = \Delta_{\mathcal{X},\mathcal{A},s}[f,f](t_3). \tag{3.29}$$

It holds

$$\Delta_0 = s\left(\mathcal{A}\left(\mathcal{X}_{00}, \cdots, \mathcal{X}_{00}, \mathcal{X}_{11}, \mathcal{X}_{00}, \mathcal{X}_{11}, \mathcal{X}_{00}, \mathcal{X}_{00}, \cdots, \mathcal{X}_{00}\right)\right), \tag{3.30}$$
$$\Delta_1 = s\left(\mathcal{A}\left(\mathcal{X}_{00}, \cdots, \mathcal{X}_{00}, \mathcal{X}_{10}, \mathcal{X}_{01}, \mathcal{X}_{10}, \mathcal{X}_{01}, \mathcal{X}_{00} \cdots, \mathcal{X}_{00}\right)\right), \tag{3.31}$$
$$\Delta_2 = s\left(\mathcal{A}\left(\mathcal{X}_{00}, \cdots, \mathcal{X}_{00}, \mathcal{X}_{10}, \mathcal{X}_{00}, \mathcal{X}_{11}, \mathcal{X}_{00}, \mathcal{X}_{01} \cdots, \mathcal{X}_{00}\right)\right). \tag{3.32}$$

Now if we assume $\mathcal{X}_{10} < \mathcal{X}_{11}$ then the condition $(C3)$ implies that $\mathcal{X}_{00} = \mathcal{X}_{10}$ and we get $\Delta_0 > \Delta_1 < \Delta_2$ due to the strict monotonicity of the aggregation function. The case $\mathcal{X}_{10} > \mathcal{X}_{11}$ implies $\mathcal{X}_{00} = \mathcal{X}_{11}$ and $\Delta_0 < \Delta_1 > \Delta_2$ thereafter. Therefore no monotonicity of the autocorrelation function can be achieved with such construction principle. □

Note that this result has only been proven for discrete signals so far. It cannot be generalized in this form to general continuous measure spaces, as the measure does not ensure any longer that the interval $[0,1)$ and $[1,2)$ (for instance) have the same measure. Therefore the comparison as it is written in this proof is no longer tractable. A general formulation of this result is still under study (it would however work for some uniform measure on a bounded domain).

A consequence of the first theorem is that combining two such similarity measures leads to another measure whose autocorrelation function does not fulfil the monotonicity property:

Corollary 1. *Let s_1, s_2, $\mathcal{A}_1$, $\mathcal{A}_2$, $\mathcal{X}_1$ and $\mathcal{X}_2$ be defined with the conditions according to Theorem 6 and let $\odot : \mathbb{R} \times \mathbb{R} \to \mathbb{R}$ be strictly monotonic of the same type in both arguments (i.e. increasing in both arguments of decreasing in both arguments). Then*

$$\Delta[f,f] := \Delta_{\mathcal{X}_1,\mathcal{A}_1,s_1}[f,f] \odot \Delta_{\mathcal{X}_2,\mathcal{A}_2,s_2}[f,f] \tag{3.33}$$

does not fulfil the monotonicity property.

Proof. It is clear whenever the strict monotonicity $\odot$ is of the same type in both arguments, as the inequalities $\Delta_0 > \Delta_1 < \Delta_2$ or $\Delta_0 < \Delta_1 > \Delta_2$ are valid for both $[\mathcal{X}, \mathcal{A}, s]$ settings. □

The two last results already give quite a lot of usual similarities, as one can see from Table 3.2 Other conditions on the scaling, component-wise operation and aggregation function can be found that would lead to the same results, as stated in [MSB11]. Particularly interesting with the Thm. 6 is its applications to all classical translation invariant kernels as used in machine learning [SS02], for instance.

Table 3.2: Examples of kernels and distance measures that follow the construction principles of Theorem 6 or Corollary 1 with summation as aggregation function

fomulae	name	remark
$\|f-g\|_p$	Minkowski distance	$\mathcal{X}(a,b)=\|a-b\|^p$, $s(x)=\sqrt[p]{x}$
$\langle f,g \rangle = \sum_i f_i \cdot g_i$	inner product	$\mathcal{X}(a,b)=a\cdot b$
$e^{-\frac{1}{\sigma}\sum_i(f_i-g_i)^2}$	Gaussian kernel	$s(x)=\exp(-x/\sigma)$
$-\sqrt{\|f-g\|^2+c^2}$	multiquadratic	$s(x)=-\sqrt{x+c^2}$
$\frac{1}{\sqrt{\|f-g\|^2+c^2}}$	inverse multiquadratic	$s(x)=(\sqrt{x+c^2})^{-1}$
$\|f-g\|^{2n}\ln(\|f-g\|)$	thin plate spline	$\ln$, x^n as scaling, $\odot(a,b)=a\cdot b$
$\langle f,g \rangle^n$, $d\in\mathbb{N}$	polynomial kernel	(Cor. 1) recursively, $\odot(a,b)=a\cdot b$
$(\langle f,g\rangle+c)^n$, $d\in\mathbb{N}$	inh. polynomial kernel	(Cor. 1) recursively, $\odot(a,b)=a\cdot b$
$\tanh(\kappa\langle x,y\rangle+\theta)$	sigmoidal kernel	$s(x)=\tanh(\kappa x+\theta)$

3.3.2.3 The case of f-divergence metrics

A similar result can be given for f-divergence measures. f-divergence measures are probabilistic measures often used in image processing tasks. We refer the reader to [PMV04, WIVA+96, VWI97] for some (non-exhaustive) references on the topic.

Let us start by recalling the basics.

f−**divergence measures** f divergence measures were introduced in the mid-60s independently in [Csi63, AS66]. A quite exhaustive review of such measures can be found in [Bas96] (in French).

Definition 12 (f-divergence measures). *Given two probability distributions absolutely continuous with respect to a reference measure μ over the set Ω and denote by p and q their probability density; moreover, let f be convex and such that $f(1)=0$, then the f divergence of P given Q is defined as*

$$D_f(P\|Q) = \int_\Omega f\left(\frac{p(x)}{q(x)}\right) q(x)\mathrm{d}\mu(x) \qquad (3.34)$$

Literally speaking, it can be understood (almost) as a generalised mean (see Def. 4.6 in the next chapter) of the quotient of both odds.

The reader should be careful when reading this last definition. Indeed so far, and in the rest of this book, f is used as a function being analysed, while in the definition of the f divergence, we have kept the "classical" notations, where P and Q (or respectively p and q) are the data observed and f is a parameter of the system (actually a parameter of the measure)

In practical examples, a signal or an image is being discretised and organised into histograms which are then normalized so that they sum up to 1. These histograms are finally interpreted as the probability distributions used in the formula. Therefore we actually work on discrete probability distributions and have the following equivalent formula:

$$D_f(P\|Q) = \sum_i f\left(\frac{p(i)}{q(i)}\right) q(i) \qquad (3.35)$$

3.3. CHARACTERISTIC PROPERTIES

Table 3.3 gives some examples of such f divergence measures often used.

Table 3.3: Some examples of f-divergence measures

Name	Formula	f	Reference
Kullback-Leibler	$D_{KL}(P\|\|Q) = \sum_i q(i) \ln\left(\frac{q(i)}{p(i)}\right)$	$t \mapsto -\ln(t)$	[Kul97]
Hellinger	$D_H(P\|\|Q) = \sum_i \left(\sqrt{p(i)} - \sqrt{q(i)}\right)^2$	$t \mapsto \left(\sqrt{t} - 1\right)^2$	[Lin91]
Bhattacharyya	$D_{BC}(P\|\|Q) = -\sum_i \sqrt{p(i)q(i)}$	$t \mapsto -\sqrt{t}$	[Bha43]

The result As stated above, a similar non monotonicity result can be proven for such $f-$ divergence measures

Theorem 7. *Let $f : [0, \infty] \to \mathbb{R} \cup \{+\infty\}$ be a strictly convex and continuous function. For two discrete sequences $A = (a_i)_{i=1}^n \in \mathcal{V}^n$ and $B = (b_i)_{i=1}^n \in \mathcal{V}^n$, $n \in \mathbb{N}$ let*

$$D_f(A\|B) = \sum_{v,w \in \mathcal{V}} P_A(v) P_B(w) f\left(\frac{P_{AB}(v,w)}{P_A(v) P_B(w)}\right) \tag{3.36}$$

where $P_{AB}(v,w)$ denotes the joint frequency of occurrences of the pair of values (v,w), and $P_A(v)$, $P_B(w)$ denote the frequencies of v, w in the corresponding sequences A and B, respectively. Then there are sequences $h : \mathbb{Z} \to \mathcal{V}$ such that $\chi : \mathbb{N} \to [0, \infty]$ given by

$$\chi_t = D_f(A_0\|A_t)$$

does not behave monotonically with respect to t, where $A_t(.) = \mathbf{1}_{\{1,\ldots,n\}}(.) \cdot h(. - t)$.

Proof. Set $\mathcal{V} = \{0, 1\}$, and define $h(.) := \sum_{j=1}^m \mathbf{1}_{\{2 \cdot j\}}(.)$ where $m \in \mathbb{N}$. Set $n = K \cdot m$ with $K \geq 3$. Then

$$P_{A_t}(0) = \frac{n-m}{n}, \quad P_{A_t}(1) = \frac{m}{n}$$

for $t \in \{0, 1, 2\}$, further

$$P_{A_0,A_0}(0,0) = \frac{n-m}{n}, \quad P_{A_0,A_0}(0,1) = 0, \quad P_{A_0,A_0}(1,0) = 0, \quad P_{A_0,A_0}(1,1) = \frac{m}{n},$$
$$P_{A_0,A_1}(0,0) = \frac{n-2m}{n}, \quad P_{A_0,A_1}(0,1) = \frac{m}{n}, \quad P_{A_0,A_1}(1,0) = \frac{m}{n}, \quad P_{A_0,A_1}(1,1) = 0,$$
$$P_{A_0,A_2}(0,0) = \frac{n-m}{n}, \quad P_{A_0,A_2}(0,1) = \frac{1}{n}, \quad P_{A_0,A_3}(1,0) = \frac{1}{n}, \quad P_{A_0,A_2}(1,1) = \frac{m-2}{n}.$$

By taking $n = K \cdot m$ into account we get

$$K^2 \chi_0(K, m) = f\left(\frac{K}{K-1}\right)(K-1)^2 + 2f(0)(K-1) + f(K)$$

$$K^2 \chi_1(K, m) = f\left(\frac{(K-2)K}{(K-1)^2}\right)(K-1)^2 + 2f\left(\frac{K}{K-1}\right)(K-1) + f(0)$$

$$K^2 \chi_2(K, m) = f\left(\frac{K}{K-1}\right)(K-1)^2 + 2f\left(\frac{K}{K-1}\frac{1}{m}\right)(K-1) + f\left(K\frac{m-2}{m}\right)$$

And we get

$$K^2 \left(\chi_0(K, m) - \chi_2(K, m)\right) = f(K) - f\left(K\frac{m-2}{m}\right) + 2(K-1)\left(f(0) - f\left(\frac{K}{K-1}\frac{1}{m}\right)\right)$$

and because of the continuity of f for all $K \geq 2$ we obtain

$$\lim_{m \to \infty} (\chi_0(K,m) - \chi_2(K,m)) = 0. \tag{3.37}$$

Now, we introduce

$$P_{A_t}(0) = \frac{n-m}{n}, \quad P_{A_t}(1) = \frac{m}{n}$$

for $t \in \{0, 1, 2\}$, further

$$\lambda_1 = \frac{(K-1)^2 - 2(K-1)}{(K-1)^2} \quad \lambda_2 = \frac{2(K-1)-1}{(K-1)^2} \quad \lambda_3 = \frac{1}{(K-1)^2}$$
$$x_1 = \frac{K}{K-1} \quad x_2 = 0 \quad x_3 = K$$

it holds

$$\sum_{i=1}^{3} \lambda_i = 1 \tag{3.38}$$

$$\sum_{i=1}^{3} \lambda_i x_i = \frac{(K-2)K}{K^2} \tag{3.39}$$

so that the strict convexity of f yields

$$f\left(\frac{(K-2)K}{K^2}\right) < \lambda_1 f(x_1) + \lambda_2 f(x_2) + \lambda_3 f(x_3)$$
$$= \frac{(K-1)^2}{(K-1)^2} f\left(\frac{K}{K-1}\right) + 2\frac{K-1}{(K-1)^2} f(0) + \frac{1}{(K-1)^2} f(K)$$
$$- 2\frac{K-1}{(K-1)^2} f\left(\frac{K}{K-1}\right) - \frac{1}{(K-1)^2} f(0) \Leftrightarrow$$
$$0 < K^2 \chi_0(K,m) - K^2 \chi_1(K,m)$$

and this is independent from m.

Therefore, combining both results together, we can find m_0 and K_0 such that $\chi_0(K_0, m_0) > K^2 \chi_1(K_0, m_0) < K^2 \chi_2(K_0, m_0)$ and therefore the monotonicity is not fulfilled. □

Chapter 4

Discrepancy correlation optimization

This section is dedicated to some approximative computations of the discrepancy norm. It is based on the convergence of the p-norms, or more precisely, the p-means towards the infinity norm. We first motivate our approaches by introducing the results for one-dimensional discrete signals and analyzing in details its computation. In particular, we see that it is possible to compute a correlation of two signals based on the discrepancy norm by a certain convolution, making the computation very efficient. We then extend these results to the continuous case. Then we see how we can compute estimates for the derivative of the discrepancy norm with respect to translations, if one wishes to make use of its monotonicity property. Finally, in a last part, those results are generalized to higher dimensional signals.

4.1 One dimensional signals

Before we give more details about our approximation, we wish to introduce some concepts which will become handy later on.

Definition 13 (Generalized mean). *Let ϕ be a continuous invertible monotone function, let $f = \{f_k\}_{k=1}^N$, we can define [KC30] a mean as*

$$M_\phi(f) = \phi^{-1}\left(\frac{\sum_{k=1}^N \phi(f_k)}{N}\right) \tag{4.1}$$

Moreover, as we would expect from a mean, we have $\min_k f_k \leq M_\phi(f) \leq \max_k f_k$. (see [HLP52] Ch II for the proof)

Now if we consider taking L^p norms as ϕ in the previous definition, this yields the following p means:

$$M_p(f) = \left(\frac{\sum_{k=1}^N |f_k|^p}{N}\right)^{1/p} \tag{4.2}$$

4.1.1 The discrete case

In [BHM10] we have introduced an approximation of the discrete discrepancy norm for one-dimensional signals based on such p-means, as stated in the following theorem, where μ_c denotes the counting measure on a finite subset of $\mathbb{R}$, and ℓ^1 represents the set of summable sequences:

Theorem 8. Let $f \in \ell^1 = L^1(\mathbb{Z}, \mu_c)$, we define

$$\gamma_p : \ell^1 \to \mathbb{R}^+$$
$$f \mapsto \gamma_p(f) := \ln\left(\frac{M_p(\chi(f))}{M_{-p}(\chi(f))}\right) \qquad (4.3)$$

where $\chi(f)(k) := \exp(\sum_{l=0}^{k} f_l)$ is the exponential of the cumulative function.
The followings hold:

$$\gamma_p(f) \leq \|f\|_D < \gamma_p(f) + \frac{2}{p}\ln(N+1) \qquad (4.4)$$

and γ_p is positive definite, i.e. $\forall f \in \mathbb{Z}^N; \gamma_p(f) \geq 0$ and equality occurs only for $f = 0$.

Proof. We define similarly χ_p as χ at the power p i.e. $\chi_p(f)(k) := \exp(p \sum_{l=0}^{k} f_l)$ We start by showing the positive-definiteness of γ_p:

$$\gamma_p(f) = 0 \quad \Leftrightarrow$$
$$M_1(\chi_p[f]) = M_{-1}(\chi_p[f]) \quad \Leftrightarrow$$
$$\forall k \in \{0, \cdots, N\},\ \chi_p[f] = C \quad \Leftrightarrow$$
$$\forall k \in \{0, \cdots, N\},\ \sum_{i=0}^{k} f_i = C' \quad \Leftrightarrow$$
$$\forall i \in \{0, \cdots, N\},\ f_i = 0$$

and γ_p is positive due to the property of the p-means being non-decreasing with respect to p.

Now let's prove the inequalities. We start by rewriting the discrepancy norm in a similar way as the γ_p function:

$$\|f\|_D = \max_{k \in \{0, \cdots, N\}} F_k + \max_{k \in \{0, \cdots, N\}} -F_k$$
$$= \ln(\max_{k \in \{0, \cdots, N\}} e^{F_k}) + \ln(\max_{k \in \{0, \cdots, N\}} e^{-F_k})$$

having written $F_k = \sum_{i=0}^{k} f_i$. Which leads to

$$\gamma_p(f) - \|f\|_D = \ln\left(\frac{\|\chi_1(f)\|_p}{\sqrt[p]{N+1}} \frac{\|\chi_{-1}(f)\|_p}{\sqrt[p]{N+1}}\right) - \ln\|\chi_1(f)\|_\infty - \ln\|\chi_{-1}(f)\|_\infty$$
$$= \ln\left(\frac{\|\chi_1(f)\|_p}{\sqrt[p]{N+1}\|\chi_1(f)\|_\infty} \frac{\|\chi_{-1}(f)\|_p}{\sqrt[p]{N+1}\|\chi_{-1}(f)\|_\infty}\right).$$

Recall that for $N \in \mathbb{N}$ and $f \in \mathbb{R}^N$ we have

$$\|f\|_\infty \leq \|f\|_p \leq \sqrt[p]{N}\|f\|_\infty,$$

hence

$$\frac{1}{\sqrt[p]{N}} \leq \frac{\|f\|_p}{\sqrt[p]{N}\|f\|_\infty} \leq 1.$$

By this, we finally obtain

$$\ln\left(\left(\frac{1}{\sqrt[p]{N+1}}\right)^2\right) \leq \gamma_p(f) - \|f\|_D \leq 0$$

which is equivalent to
$$\gamma_p(f) \leq \|f\|_D \leq \gamma_p(f) + \frac{2}{p}\ln(N+1).$$

□

Finally, for practical applications, we can choose a p as small as possible for the approximation according to the following corollary

Corollary 2 (Choice of p).

$$\forall \varepsilon > 0, \forall f \in \ell^1, \forall p \in \mathbb{R}, p \geq p^* := \frac{2}{\varepsilon}\ln(N+1) \Rightarrow |\gamma_p(f) - \|f\|_D| \leq \varepsilon \qquad (4.5)$$

Proof. According to the previous theorem, we have $|\gamma_p(f) - \|f\|_D| < \frac{2}{p}\ln(N+1)$ so the corollary holds true whenever $\frac{2}{p}\ln(N+1) \leq \varepsilon \Leftrightarrow p \geq p_0$.   □

This last result is interesting in order to avoid overflow or numerical problems due to too fine approximations. Conversely, using a time series of 1000 samples and an approximation at the power $p = 8$, the previous result ensures that the absolute error introduced can not be higher than 0.86256, independently from any characteristics of f.

4.1.2 Alignment by convolution

Another advantage of our approximation is its fast computation when one wants to align two time series. In such cases, we are given two signals or functions f and g and we wish to find the optimal translation parameter that minimizes the discrepancy of the difference: $\Delta_D[g, f](t) = \|g - \tau_t f\|_D$. Such correlation computations are particularly appreciated when a convolution can be computed. This is the case of our approximation according to the following theorem [BHM10]:

Theorem 9 (Alignment by convolution). *Given two discrete signals f and g their misalignment function can be approximated by summing up two convolutions.*

Proof. First let us introduce for simplicity $h : \mathbb{Z} \to \mathbb{R}^+, t \mapsto \gamma_p(\tau_t f - g)$. First, we can remark that $\sum_{m=1}^{n} f_{m+t} = F(n+t) - F(t)$, with F corresponding to the cumulative function of f. It holds (G being the cumulative function g)

$$\begin{aligned}
h(t) &= \frac{1}{p}\ln\left(\frac{\sum_{m=0}^{n} e^{p(F_{m+t} - F_t - G_m)}}{N+1}\right) + \frac{1}{p}\ln\left(\frac{\sum_{m=0}^{n} e^{-p(F_{m+t} - F_t - G_m)}}{N+1}\right) \\
&= \frac{1}{p}\ln\left(\frac{\sum_{m=0}^{n} e^{p(F_{m+t} - G_m)}}{N+1}\right) + \frac{1}{p}\ln\left(\frac{\sum_{m=0}^{n} e^{-p(F_{m+t} - G_m)}}{N+1}\right) \\
&= \frac{1}{p}\ln\left(\frac{\sum_{m=0}^{n} e^{p(F_{m+t})} \cdot e^{p(-G_m)}}{N+1}\right) + \frac{1}{p}\ln\left(\frac{\sum_{m=0}^{n} e^{-p(F_{m+t})} \cdot e^{-p(-G_m)}}{N+1}\right) \\
&= \frac{1}{p}\ln\left(\frac{\chi_p(f) * \chi_p(-g)}{N+1}\right) + \frac{1}{p}\ln\left(\frac{\chi_p(-f) * \chi_p(g)}{N+1}\right)
\end{aligned}$$

□

Note that this last theorem allows to find an optimal alignment of two time series of size N in a complexity of $\mathcal{O}(N\log(N))$ (actually with a factor of 2 due to the 2 convolutions).

4.1.3 The continuous case

Now that we have seen how practical and how close the approximation can be, we want to extend this to continuous measurable functions [BB13].

From now on we go back to the setting of measure space on $\mathbb{R}$: $(\mathbb{R}, \Sigma, \mu)$ with μ being a finite measure.

In the same way, we can introduce the continuous counter-part of the generalized means

$$M_\phi(f) = \phi^{-1}\left(\frac{\int_\mathbb{R} \phi(f(x))d\mu(x)}{\int_\mathbb{R} d\mu}\right) \quad (4.6)$$

And the following bounding property still holds $\inf_{x\in\mathbb{R}} f(x) \leq M_\phi(f) \leq \max_{x\in\mathbb{R}} f(x)$ (where inf and sup are understood as their essential definitions: $\mu\{x : f(x) > \sup f\} = 0$). We also re-define the p-means M_p in the same way.

The proposed approximation is reintroduced in the same way (with capital letter for the continuous case):

$$\Gamma_p : L^1(\mathbb{R}, \mu) \to \mathbb{R}^+$$
$$f \mapsto \Gamma_p(f) := \ln\left(\frac{M_p(\chi(f))}{M_{-p}(\chi(f))}\right) \quad (4.7)$$

where χ is defined as the exponential of the cumulative function.

The following theorem holds:

Theorem 10 (Approximation of the discrepancy norm for 1D continuous functions). Γ_p *defines an approximation of the discrepancy norm in the sense that*

i) $\forall f \in L(\mathbb{R}, \mu), \Gamma_p(f) \xrightarrow[p\to\infty]{} \|f\|_D$

ii) $\Gamma_p(f) \leq \|f\|_D$

iii) $\forall \varepsilon > 0, \exists p^* : \forall p \in \mathbb{R}, p \geq p^* \Rightarrow |\Gamma_p(f) - \|f\|_D| \leq \varepsilon$

Proof. As we have seen, we can define

$$\chi : L(\mathbb{R}, \mu) \to L(\mathbb{R}, \mu)$$
$$f \mapsto \chi(f) : \chi(f)(x) = e^{\int_{-\infty}^x f(t)d\mu(t)} \quad (4.8)$$

Let $f \in L(\mathbb{R}, \mu)$, we need to compute $d_p(f) := \|f\|_D - \Gamma_p(f)$. First we rewrite $\|f\|_D$ another way, so that it gets comparable to Γ_p.

$$\begin{aligned}
\|f\|_D &= \sup_b \int_{-\infty}^b f d\mu - \inf_a \int_{-\infty}^a f d\mu \\
&= \sup_b \int_{-\infty}^b f d\mu + \sup_a -\int_{-\infty}^a f d\mu \\
&= \sup_b \ln(\chi(f)(b)) + \sup_a \ln(\chi(-f)(a)) \\
&= \ln\left(\sup_b \chi(f)(b)\right) + \ln\left(\sup_a \chi(-f)(a)\right), \text{ as } \ln \text{ is monotonic.} \\
&= \ln(\|\chi(f)\|_\infty \cdot \|\chi(-f)\|_\infty) \quad (4.9)
\end{aligned}$$

4.1. ONE DIMENSIONAL SIGNALS

Plugging this into the expression of d_p, and noting that $\Gamma_p(f) = \ln\left(\frac{\|\chi(f)\|_p}{\mu(\mathbb{R})^{1/p}} \cdot \frac{\|\chi(-f)\|_p}{\mu(\mathbb{R})^{1/p}}\right)$ (clear, left to the reader) we get:

$$d_p(f) = \ln\left(\frac{\|\chi(f)\|_\infty \mu(\mathbb{R})^{1/p}}{\|\chi(f)\|_p} \cdot \frac{\|\chi(-f)\|_\infty \mu(\mathbb{R})^{1/p}}{\|\chi(-f)\|_p}\right) \tag{4.10}$$

We make use of the following lemma:

Lemma 1 (p–norm approximation of the ∞ norm). *Let f be in $L^q \cap L^\infty$ for a certain $q \in \mathbb{R}$. It holds:*

$$\|f\|_p \xrightarrow[p\to\infty]{} \|f\|_\infty \tag{4.11}$$

A proof of this lemma can be found in Ch.III, Theorem 14F. of [Loo53].

If we apply the previous lemma to $\chi(f)$, we get that $\frac{\|\chi(f)\|_\infty}{\|\chi(f)\|_p} \xrightarrow[p\to\infty]{} 1$ and $\mu(\mathbb{R})^{1/p} \xrightarrow[p\to\infty]{} 1$. The same holds for $\chi(-f)$ and therefore we have $d_p(f) \xrightarrow[p\to\infty]{} 0$ which proves the first point of the theorem.

For the second point, we make use of the following inequality:

$$\forall f \in L^p(\mathbb{R},\mu) \cap L^q(\mathbb{R},\mu), 1 \leq p \leq q \leq \infty, \|f\|_p \leq \mu(\mathbb{R})^{1/p-1/q}\|f\|_q \tag{4.12}$$

If we apply this inequality to $\chi(f)$ and $\chi(-f)$ again, we get:

$$\frac{\mu(\mathbb{R})^{(1/p)}\|\chi(f)\|_\infty}{\|\chi(f)\|_p} \geq 1 \qquad \frac{\mu(\mathbb{R})^{(1/p)}\|\chi(-f)\|_\infty}{\|\chi(-f)\|_p} \geq 1$$

$$\ln\left(\frac{\mu(\mathbb{R})^{(1/p)}\|\chi(f)\|_\infty}{\|\chi(f)\|_p}\right) \geq 0 \qquad \ln\left(\frac{\mu(\mathbb{R})^{(1/p)}\|\chi(-f)\|_\infty}{\|\chi(-f)\|_p}\right) \geq 0 \tag{4.13}$$

And therefore, we get the second results by summing up both contributions.

For computing a minimum rank of convergence within ε, we make use of the following lemma:

Lemma 2 (Eventual convergence of L^p means towards ∞ norm). *Let $\varepsilon' > 0$ and $g \in L^q \cap L^\infty$ for a certain $q \in \mathbb{R}$. For $\varepsilon' < 1, 0 < t < \varepsilon'$, let $p_0 = p_0(t, \varepsilon', g) = \frac{\ln\left(\frac{\mu(E_g(t))}{\mu(\mathbb{R})}\right)}{\ln\left(\frac{1-\varepsilon'}{1-t}\right)}$, with $E_g(t) = \{x : |g(x)| > \|g\|_\infty(1-t)\}$. It holds*

$$\forall p \geq \max\{q, p_0\}, \frac{M_p(g)}{\|g\|_\infty} \geq 1 - \varepsilon' \tag{4.14}$$

Proof. The convergence is clear due to the previous lemma and inequality 4.12. Now fix $\varepsilon' > 0$ and let $1 > \varepsilon' \geq t = \alpha\varepsilon' > 0$ with $0 < \alpha < 1$ and consider the set $E_g(t) = \{x : |g(x)| > \|g\|_\infty(1-t)\}$, it holds:

$$M_p(g) = \left(\frac{\int_\mathbb{R} |g|^p d\mu}{\mu(\mathbb{R})}\right)^{1/p}$$
$$= \left(\frac{\int_{E_g(t)} |g|^p d\mu + \int_{E_g(t)^c} |g|^p d\mu}{\mu(\mathbb{R})}\right)^{1/p}$$
$$\geq \left(\frac{\mu(E_g(t))}{\mu(\mathbb{R})}\right)^{1/p} (\|g\|_\infty(1-t)) \tag{4.15}$$

Based on this, we are looking for a lower-bound p_0 of p such that $\frac{M_p(g)}{\|g\|_\infty} \geq 1 - \varepsilon'$:

$$M_p(g) \geq \left(\frac{\mu(E_g(t))}{\mu(\mathbb{R})}\right)^{1/p} (\|g\|_\infty(1-t)) \geq \|g\|_\infty(1-\varepsilon') \quad \Leftrightarrow$$

$$\left(\frac{\mu(E_g(t))}{\mu(\mathbb{R})}\right)^{1/p} \geq \frac{1-\varepsilon'}{1-t} \quad \Leftrightarrow$$

$$\frac{1}{p}\ln\left(\frac{\mu(E_g(t))}{\mu(\mathbb{R})}\right) \geq \ln\left(\frac{1-\varepsilon'}{1-t}\right), \frac{1-\varepsilon'}{1-t} < 1 \quad \Leftrightarrow$$

$$p \geq \frac{\ln\left(\frac{\mu(E_g(t))}{\mu(\mathbb{R})}\right)}{\ln\left(\frac{1-\varepsilon'}{1-t}\right)} \quad (4.16)$$

Last expression holds for any $0 < t < \varepsilon'$. For instance, for $t = \varepsilon'/2$ (or $\alpha = 1/2$)

$$\forall p \geq p_0 = \frac{\ln\left(\frac{\mu(E_g(\varepsilon'/2))}{\mu(\mathbb{R})}\right)}{\ln\left(2\frac{1-\varepsilon'}{2-\varepsilon'}\right)}, M_p(g) \geq \|g\|_\infty(1-\varepsilon') \quad (4.17)$$

This finishes the proof of the lemma. □

Now we apply this lemma to both $\chi(f)$ and $\chi(-f)$ and choosing $\varepsilon' = 1 - e^{-\varepsilon/2}$ and we therefore get two constants $p^+ = p_0(t, \varepsilon', \chi(f))$ and $p^- = p_0(t, \varepsilon', \chi(-f))$ and we have $\forall p \geq p^* := \max\{p^+, p^-\}, \Gamma_p(f) \geq \|f\|_D - \varepsilon$.

This finishes the proof of the theorem. □

Remarks This lower bound depends strongly on the function f through the set $E_f(t)$; this is due to the non-equivalence of norms in the continuous case (in the discrete case, the bounds given for the theorem were crisp and uniform).

In the proof we have made use of an upper level set $E_f(t)$ and have proven that we get an upper bound for $t = \varepsilon/2$. It would work for $t = \varepsilon/4$ too and more generally for any $0 < t < \varepsilon$ (or equivalently $0 < \alpha < 1$). We have no idea at the moment what would be the influence of this on the estimated p^*.

Finding a nice p^* is a crucial problem for practical applications. Indeed because we are taking the pth power of an exponential, the risk to encounter numerical instability is high. Therefore we need to ensure a good approximation but keeping the power p as small as possible.

Example of lower bounds For illustration purposes let us see this effect on some toy examples.

We consider three functions:

$$f_1 : [0, 5] \to \mathbb{R}$$
$$x \mapsto 1 \text{ if } x \in [1, 2) \cap [3, 4), \ 0 \text{ elsewhere} \quad (4.18)$$
$$f_2 : [-10, 10] \to \mathbb{R}$$
$$x \mapsto \sin(x) \quad (4.19)$$
$$f_3 : [-10, 10] \to \mathbb{R}$$
$$x \mapsto a \cdot x + \sin(x) \quad (4.20)$$

4.1. ONE DIMENSIONAL SIGNALS

where we have chosen a to in the order of $1/10$. In our case, we consider a measure as a combination of uniformly distributed Diracs (every 0.02 for f_1 and every 0.01 for the two others.)
f_1, f_2, f_3 are illustrated in Fig. 4.1.

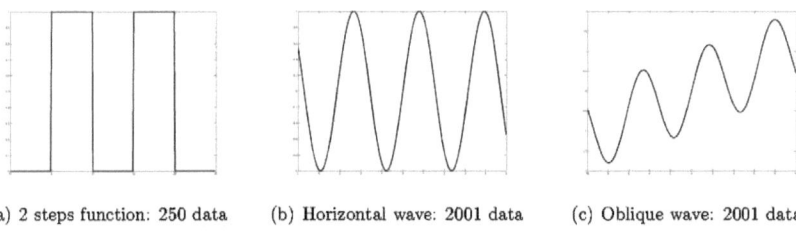

(a) 2 steps function: 250 data (b) Horizontal wave: 2001 data (c) Oblique wave: 2001 data

Figure 4.1: Our dataset of 1 dimensional toy functions used as illustrative examples.

We start by analyzing the convergence of the proposed p−norm approximations towards the discrepancy norm. The two first statements of Theorem 10 tell that Γ_p gets to $\|\cdot\|_D$ from below with p increasing. Figs. 4.2 depict this behaviour. One sees that the values of Γ_p functions increase drastically with p close to 0 and tends to stabilize to $\|\cdot\|_D$ very soon.

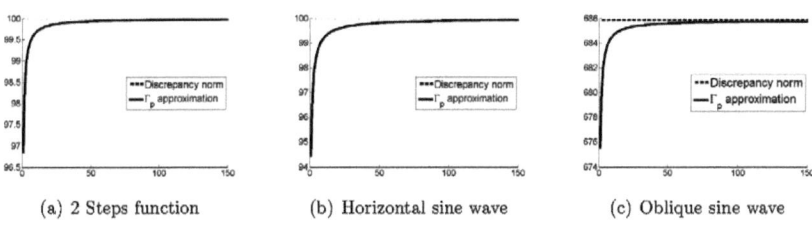

(a) 2 Steps function (b) Horizontal sine wave (c) Oblique sine wave

Figure 4.2: Convergence of the Γ_p functions towards the discrepancy norm. The horizontal axis corresponds to changing values of p while the vertical one is the output of the Γ_p and discrepancy norm functions.

The next point we need to have a look at is the reliability of our estimator for the value of p^*. While this value is clearly not optimal (we have left potentially a lot of the function away when deriving the estimation), we would like it to be relatively small anyway. As this is the value we are going to use afterwards for our tests, we need it to be small enough, in order not to get into machine overflow.

We have gathered results based on our three toy functions. For all of them, we have noted the p^* defined in Theorem 10, and the smallest integer value, denoted by $\overline{p^*}$ and considered as an integer for simplicity reasons, for which the approximation up to ε holds. Moreover, as it can be seen in the proof of the theorem, one can tune the threshold used for the decomposition of the function. This threshold has also been analyzed for some particular values of α.

The results are summarized in Table. 4.1.

It comes out that we indeed get close approximation of the discrepancy norm of a function by means of p−norms approximation. The estimator of a good p^* is however strongly dependent on the choice of the threshold. However first tests tend to show that when given an α value for the threshold in the order of 0.1 or 0.2 yield stable and reliable results.

Table 4.1: Examples of approximation and lower bound estimations on some toy functions.

Function	ε	α	p^*	$\Gamma_{p^*}(f)$	$\|f\|_D$	$\overline{p^*}$
2 steps	0.1	1/2	100.77	99.97	100	32
	5	0.2	2.68	98.81	100	1
Horizontal sines	0.1	1/4	257.73	99.96	100.00	101
	0.1	1/2	348.72	99.97	100.00	101
	0.1	3/4	666.56	99.98	100.00	101
	0.05	1/2	767.06	99.98	100.00	212
	2	1/2	13.45	99.39	100.00	4
Oblique sines	1	1/2	105.99	685.72	685.85	13
	2	1/4	44.64	685.55	685.85	7

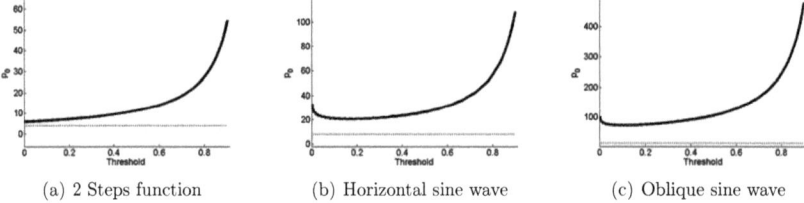

(a) 2 Steps function   (b) Horizontal sine wave   (c) Oblique sine wave

Figure 4.3: Computed values of the p^* estimator. The continuous curves show the estimated p^* while the the dotted straight line corresponds to the $\overline{p^*}$ as described in the Table 4.1

Figs. 4.3 show the behaviour of the estimation of p^* for different thresholding values. For the rest of this work we use in our applications an α value for thresholding of 0.2 which seems to lie in a rather stable area and which achieves low values of p^*. This stability is actually not seen in the first example (Fig. 4.3(a)). This is due to the fact that the function is actually binary which means that the term $\mu(E_f(t))$ does not change with higher threshold values which leads to a continuous decrease of the estimated p^*.

4.1.4 Derivation of the autocorrelation function

In this section we introduce a way of optimising the discrepancy norm based correlation function. Given a function f and a reference g, we want to find the optimal translation parameter t^* such that $\|f_t - g\|_D$ is minimal (for some t), where $\forall t \in \mathbb{R}, \tau_t f(x) = f(x - t)$. We consider the following optimization problem:

$$t^* = \underset{t \in \mathbb{R}}{\operatorname{argmin}} \|\tau_t f - g\|_D \qquad (4.21)$$

We would like to optimize the objective function $J(t) = \|\tau_t f - g\|_D$ making use of the monotonicity property of the discrepancy norm when facing misaligned functions. However, due to its definition with max and min, the discrepancy norm is only almost everywhere differentiable. Moreover, as it can be seen from Fig. 4.4, the objective function shows some kind of plateau on some critical cases which yield any gradient based method to fail in finding a correct solution. Therefore, we want to make use of the previous approximation to compute an approximated gradient to the objective function. As we

4.1. ONE DIMENSIONAL SIGNALS

see, we can derive formula based on a scalar product for efficient computations.

Theorem 11 (Derivation of the discrepancy correlation). *The discrepancy correlation function's derivative $J(t)$ can be approximated in the following way:*

$$\frac{dJ}{dt}(t) \approx \left\langle \int_{x=-\infty} f'(x-t)\,d\mu(x), \left(\frac{\chi(-\tau_t f + g)}{\|\chi(-\tau_t f + g)\|_p}\right)^p - \left(\frac{\chi(\tau_t f - g)}{\|\chi(\tau_t f - g)\|_p}\right)^p \right\rangle \quad (4.22)$$

Proof. We need to compute the derivative of the discrepancy norm of a difference. As it has been seen in the previous section, it can be approximated by means of L^p norms and we get

$$\frac{dJ}{dt}(t) \approx \frac{\partial \Gamma_p}{\partial t}(\tau_t f - g) = \frac{\partial}{\partial t}\left(\frac{1}{p}\ln\left(\|\chi(\tau_t f - g)\|_p^p\right) + \frac{1}{p}\ln\left(\|\chi(-\tau_t f + g)\|_p^p\right) + C\right) \quad (4.23)$$

$$\approx \frac{1}{p}\frac{\partial}{\partial t}\left(\Gamma_p^{(+)}(\tau_t f - g) + \Gamma_p^{(-)}(\tau_t f - g)\right) \quad (4.24)$$

Where we have defined $\Gamma_p^{(+)}(h) := \ln\left(\|\chi(h)\|_p^p\right) := \Gamma_p^{(-)}(-h)$,

$$\frac{\partial \Gamma_p^{(\pm)}}{\partial t}(\tau_t f - g) = \frac{1}{\|\chi(\pm(\tau_t f - g))\|_p^p}\frac{\partial}{\partial t}\|\chi(\pm(\tau_t f - g))\|_p^p, \text{ with } F_p(h) := \|\chi(h)\|_p^p \quad (4.25)$$

$$\frac{\partial F_p}{\partial t}(\pm(\tau_t f - g)) = \frac{\partial}{\partial t}\int_{\mathbb{R}}|\chi(\pm(\tau_t f - g))|^p\,d\mu = \frac{\partial}{\partial t}\int_{\mathbb{R}}e^{\pm(F_t(x)-G(x))}\,d\mu(x) \quad (4.26)$$

$$= \int_{\mathbb{R}}\frac{\partial}{\partial t}e^{\pm(F_t(x)-G(x))}\,d\mu(x) \quad (4.27)$$

$$= -\pm\int_{\mathbb{R}}\left(\int_{x=-\infty} f'(x-t)\,d\mu(x)\right)\chi(\pm(\tau_t f - g))^p\,d\mu \quad (4.28)$$

and we finally get

$$\frac{\partial \Gamma_p}{\partial t}(\tau_t f - g) = \frac{-1}{\|\chi(\tau_t f - g)\|_p^p}\int_{\mathbb{R}}\left(\int_{x=-\infty} f'(x-t)\,d\mu(x)\right)\chi(\tau_t f - g)^p\,d\mu$$

$$+ \frac{1}{\|\chi(-\tau_t f + g)\|_p^p}\int_{\mathbb{R}}\left(\int_{x=-\infty} f'(x-t)\,d\mu(x)\right)\chi(-\tau_t f + g)^p\,d\mu \quad (4.29)$$

$$\frac{\partial \Gamma_p}{\partial t}(\tau_t f - g) = \left\langle \int_{x=-\infty} f'(x-t)\,d\mu(x), \left(\frac{\chi(-\tau_t f + g)}{\|\chi(-\tau_t f + g)\|_p}\right)^p - \left(\frac{\chi(\tau_t f - g)}{\|\chi(\tau_t f - g)\|_p}\right)^p \right\rangle \quad (4.30)$$

and this finishes the proof. □

While this formula seems complicated, we can see that on real discrete signals applications, it can be simplified a little

Corollary 3 (Derivation for discrete uniform measure). *Assume μ is a uniform discrete measure on a bounded domain. Then the derivation of the discrepancy autocorrelation function can be approximated by*

$$\frac{dJ}{dt}(t) \approx \left\langle \tau_t f, \left(\frac{\chi(-\tau_t f + g)}{\|\chi(-\tau_t f + g)\|_p}\right)^p - \left(\frac{\chi(\tau_t f - g)}{\|\chi(\tau_t f - g)\|_p}\right)^p \right\rangle \quad (4.31)$$

or equivalently:

$$\frac{\mathrm{d}J}{\mathrm{d}t}(t) \approx \sum_{i=1}^{N} \tau_t f(i) \left(\left(\frac{\chi(-\tau_t f + g)(i)}{\|\chi(-\tau_t f + g)\|_p} \right)^p - \left(\frac{\chi(\tau_t f - g)(i)}{\|\chi(\tau_t f - g)\|_p} \right)^p \right) \quad (4.32)$$

where N denotes the size of the discrete finite signal (i.e. vector) we consider.

As stated earlier let us see on the toy example two steps function how the discrepancy norm can be problematic for local gradient based optimization algorithms. If we consider the autocorrelation function of the 2 steps example based on the discrepancy norm as well as on the Γ_p functions (see Fig. 4.4), it appears that even on really easy mock example, the discrepancy norm does not appear to be well suited for optimization based procedure.

However, using the above introduced approximations allows us to have an appropriate idea of what the discrepancy objective function looks like while keeping differentiable functions and avoiding plateaus which causes the derivative to vanish.

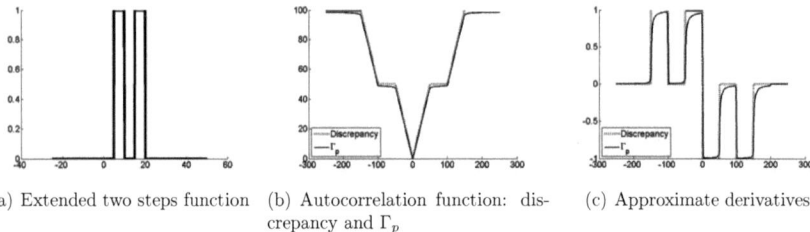

(a) Extended two steps function   (b) Autocorrelation function: discrepancy and Γ_p (c) Approximate derivatives

Figure 4.4: This figure shows how a gradient descent based optimization algorithm might fail when trying to locally optimize the discrepancy correlation function. Indeed the correlation function (second figure) shows some plateau where the derivative (illustrated on the third figure) is 0. On the other hand, using an appropriate p−norm approximation allows to overcome this effect while keeping a really close objective function.

Now let us see how it behaves on some concrete examples. If we consider the horizontal wave function f and take it as a pattern we would like to align into a bigger pattern g, the discrepancy norm and its approximation(s) should ideally behave similarly when translating f along g. As toy examples, we consider two cases for g. The first one extends the f pattern only by padding 0s outside the domain of f. The second case mirrors the function f at its borders. These examples are illustrated in Figs. 4.5(a) to 4.5(c). The last figure illustrates the case when the frequency of the wave is twice as big.

For a given function, we have computed the discrepancies between the local window of g and our pattern f. The same has been done for Γ_p where p has been chosen according to Theorem 10 using $\varepsilon = 0.1$ and a threshold of 0.2. The values are illustrated for displacement in the range of $[-N, N]$ for a uniform Dirac comb as measure in Figs. 4.5(d) to 4.5(f). It appears that the approximation overlap well with the original discrepancy autocorrelation function. Moreover, as we are aiming at minimizing this correlation function, we can remark that the different local minima and global minimum are located at the same places.

Finally, the last row of Fig. 4.5 shows two approximations of the discrepancy correlation. One is concerned with the analytical derivative of the Γ_p function over the shift, and the other one is computed

4.2. HIGHER-DIMENSIONAL SIGNALS

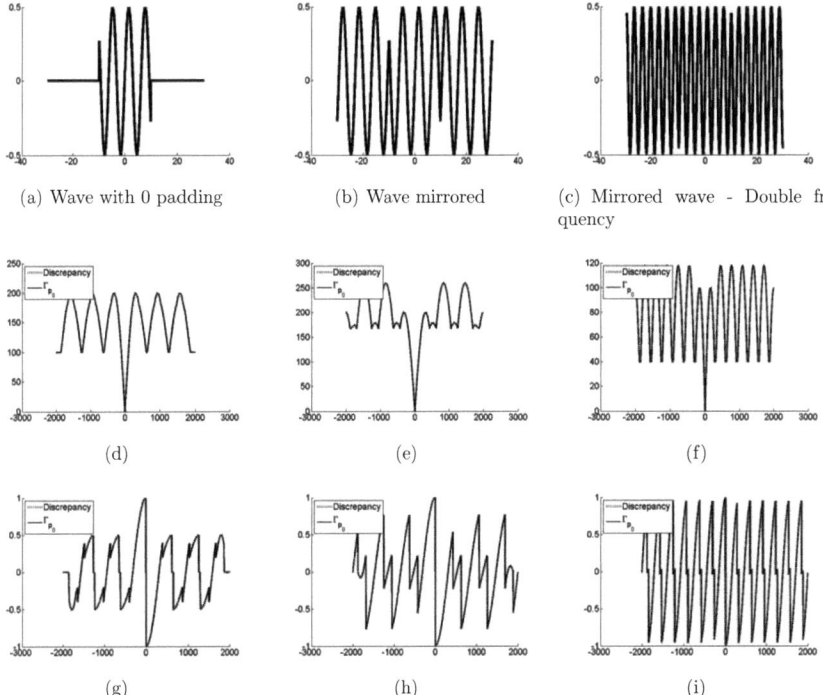

Figure 4.5: Sample sine wave with their discrepancy autocorrelation functions and their approximate derivatives. The first row shows the input signal extended with 0 padding (first column) and mirroring the wave (2nd and 3rd columns). 2nd row shows the autocorrelation function based on the discrepancy norm and its approximation. Last row shows the approximated derivative together with the finite difference of the discrepancy autocorrelation.

as a finite difference of the discrepancy correlation: $\frac{\partial}{\partial t}\left(\Delta_{\|\cdot\|_D}[f](t)\right) \approx \Delta_{\|\cdot\|_D}[f](t+1) - \Delta_{\|\cdot\|_D}[f](t)$. We see that not only is the approximated derivative fast to compute (only p-norms computations and a scalar product) we also get results pretty close to what numerical differentiation would give. Finally we can see that this approximation is suitable for any gradient based local optimization methods.

4.2 Higher-dimensional signals

4.2.1 Approximation formula

Now that the theory is clear for one-dimensional functions, we would like to extend it to handle multivariate functions. According to [Mos11] and as we have detailed in the previous chapter, there are different ways of extending the discrepancy norm to higher dimensional function but we will stick to the following:

Definition 14 (Multivariate discrepancy norm). *The discrepancy norm for multivariate functions can be defined by extending Eq. 3.16 to multidimensional integration. Let $\iota \in \{-1, 1\}^n$ and for simplicity* reason, let us define, for $s \in \mathbb{R}^n$, $[s[:=]-\infty, s_1[\times \cdots \times]-\infty, s_n[$. For a given ι, we denote by $\iota \cdot [s[$, the set composed of a product of integrals: $\iota \cdot [s[:=]-\iota_1\infty, \iota_1 s_1[\times \cdots \times]-\iota_n\infty, \iota_n s_n[$, where the intervals might be flipped. Then we have:

$$\|f\|_I^{(n)} := \max_{\iota \in \{-1,1\}^n} \left\{ \sup_{s \in \mathbb{R}^n} \int_{\iota \cdot [s[} f \, d\mu - \inf_{s \in \mathbb{R}^n} \int_{\iota \cdot [s[} f \, d\mu \right\} \quad (4.33)$$

And this equation can be numerically implemented by means of integral images [Cro84, VJ01]. For simplicity, we define the following notation:

$$D^{(\iota)}(f) := \sup_{s \in \mathbb{R}^n} \int_{\iota \cdot [s[} f \, d\mu - \inf_{s \in \mathbb{R}^n} \int_{\iota \cdot [s[} f \, d\mu \quad (4.34)$$

It defines the discrepancy of a function given a direction of integration (by direction, we mean, from which corner of the space we start the integral image.)

Therefore we can equivalently write $\|f\|_I^{(n)} = \max_{1 \leq i \leq 2^n} \{D^{(\iota^{(i)})}(f)\}$. We also define an approximation function through p-norms:

$$\Gamma_p^{(\iota)}(f) := \ln\left(M_p(\chi^{(\iota)}(f)) \cdot M_p(\chi^{(\iota)}(-f))\right) = \ln\left(\frac{\|\chi^{(\iota)}(f)\|_p}{\mu(\mathbb{R}^n)^{(1/p)}} \cdot \frac{\|\chi^{(\iota)}(-f)\|_p}{\mu(\mathbb{R}^n)^{(1/p)}}\right) \quad (4.35)$$

As in the previous sections, we have the following approximation theorem:

Theorem 12 (Approximation of the directional discrepancy norm of multivariate functions). $\Gamma_p^{(\iota)}$ defines an approximation of the directional discrepancy norm in the sense that

i) $\forall f \in L(\mathbb{R}^n; \mu), \Gamma_p^{(\iota)}(f) \xrightarrow[p \to \infty]{} D^{(\iota)}(f)$

ii) $\Gamma_p(f) \leq D^{(\iota)}(f)$

iii) $\forall \varepsilon > 0, \exists p^{(d,\iota)} : \forall p \in \mathbb{R}, p \geq p^{(d,\iota)} \Rightarrow |\Gamma_p^{(\iota)}(f) - D^{(\iota)}(f)| \leq \varepsilon$

Proof. The proof is the same as for Theorem 10, for continuous univariate functions. However, for the last point, p^* has to be changed according to the dimension;

$$p^{(d,\iota)} := \max\left\{ \frac{\ln \frac{\mu(E_{\chi^{(\iota)}(f)}(t))}{\mu(\mathbb{R}^n)}}{\ln \frac{1-\varepsilon'}{1-t}}, \frac{\ln \frac{\mu(E_{\chi^{(\iota)}(-f)}(t))}{\mu(\mathbb{R}^n)}}{\ln \frac{1-\varepsilon'}{1-t}} \right\}$$

with $\varepsilon' := 1 - e^{\varepsilon/2}$ □

However, the problem is not solved yet. Indeed, we are able to smoothly approximate the directional discrepancies, but we still have to combine them by taking the max over all 2^n possible directions. Fortunately, this max over a finite set can be once again approximated with a reasonable error with q−norms:

$$\|f\|_D = \max_{\iota \in \{-1,1\}^n} D^{(\iota)}(f) \approx \|\vec{D}(f)\|_q \quad (4.36)$$

where $\vec{D} = \left[D^{(\iota^{(1)})}, \cdots, D^{(\iota^{(2^n)})}\right]^T$ denotes the column vector composed with all directional discrepancies.

4.2. HIGHER-DIMENSIONAL SIGNALS

Finally we can combine both $p-q$−norm approximations:

$$\Gamma_{p,q}^{(n)}(f) := \left(\frac{\sum_{\iota \in \{-1,1\}^n} \Gamma_p^{(\iota)q}(f)}{2^n}\right)^{1/q} = M_q(\overrightarrow{\Gamma_p}) \tag{4.37}$$

where we have $\overrightarrow{\Gamma_p} = \left[\Gamma_p^{(\iota^{(1)})}, \cdots, \Gamma_p^{(\iota^{(2^n)})}\right]^T$

Corollary 4 (Convergence of $\Gamma_{p,q}$). *Given a function $f \in L(\mathbb{R}^n, \mathbb{R}, \mu)$, the following holds:*

$$\forall \varepsilon > 0, \exists \mathbf{p}^{(n)} := (p^{(n)}, q^{(n)}) : \forall \mathbf{p} \geq \mathbf{p}_0, \left|\Gamma_{p,q}^{(n)}(f) - \|f\|_I^{(n)}\right| \leq \varepsilon \tag{4.38}$$

Note that $\mathbf{p} = (p,q) \geq \mathbf{p}^{(n)}$ reads $p \geq p^{(n)}$ and $q \geq q^{(n)}$.

Proof. Let $0 < \lambda < 1$ and $\varepsilon > 0$.

We need to estimate the quantity $\delta_{p,q} := \left|\|f\|_I^{(n)} - \Gamma_{p,q}^{(n)}(f)\right|$.

$$\delta_{p,q} = \|f\|_I^{(n)} - M_q(\overrightarrow{D}(f)) + M_q(\overrightarrow{D}(f)) - \Gamma_{p,q}^{(n)}(f) =: d_1 + d_2 \tag{4.39}$$

Let us start with the first component

$$d_1 = \|f\|_I^{(n)} - M_q(\overrightarrow{D}(f)) = \|\overrightarrow{D}(f)\|_\infty - M_q(\overrightarrow{D}(f)) \tag{4.40}$$

this last difference actually lies in a finite-dimensional vector space (2^n dimensions) and we have the following norm comparisons:

$$\|\overrightarrow{D}(f)\|_\infty \leq \|\overrightarrow{D}(f)\|_q \leq 2^{n/q}\|\overrightarrow{D}(f)\|_\infty \tag{4.41}$$

which means that for q big enough ($q \geq q_0$), we have $\left|\|\overrightarrow{D}(f)\|_q - \|\overrightarrow{D}(f)\|_\infty\right| < (1-\lambda)\varepsilon$. Indeed

$$\|\overrightarrow{D}(f)\|_\infty \leq \|\overrightarrow{D}(f)\|_q \leq 2^{n/q}\|\overrightarrow{D}(f)\|_\infty$$

$$\frac{\|\overrightarrow{D}(f)\|_\infty}{2^{n/q}} \leq M_q(\overrightarrow{D}(f)) \leq \|\overrightarrow{D}(f)\|_\infty$$

$$\frac{\|\overrightarrow{D}(f)\|_\infty}{2^{n/q}} - \|\overrightarrow{D}(f)\|_\infty \leq M_q(\overrightarrow{D}(f)) - \|\overrightarrow{D}(f)\|_\infty \leq 0$$

whence for $-(1-\lambda)\varepsilon \leq \|\overrightarrow{D}(f)\|_\infty \left(2^{-n/q} - 1\right)$ we get the results.

$$1 - \frac{(1-\lambda)\varepsilon}{\|\overrightarrow{D}(f)\|_\infty} \leq 2^{-n/q}$$

$$\ln\left(\frac{\|\overrightarrow{D}(f)\|_\infty - (1-\lambda)\varepsilon}{\|\overrightarrow{D}(f)\|_\infty}\right) \leq -\frac{n}{q}\ln(2)$$

$$\frac{n\ln(2)}{\ln\left(\frac{\|\overrightarrow{D}(f)\|_\infty}{\|\overrightarrow{D}(f)\|_\infty - (1-\lambda)\varepsilon}\right)} \leq q$$

And we set $q_0 := n\dfrac{\ln 2}{\ln\left(\frac{\|f\|_I^{(n)}}{\|f\|_I^{(n)} - (1-\lambda)\varepsilon}\right)}$

We now need to compute an estimate for the second part fixing $q = q_0$, and try to get $d_2 \leq \lambda\varepsilon$ and so would be the corollary proven:

$$d_2 = M_q(\vec{D}(f)) - \Gamma_{p,q}^{(n)}(f) = M_q(\vec{D}(f)) - M_q(\vec{\Gamma_p}(f)) \leq \lambda\varepsilon \Leftrightarrow$$

$$= \frac{\|\vec{\Gamma_p}(f)\|_q - \|\vec{D}(f)\|_q}{2^{n/q}} \leq \lambda\varepsilon$$

Now applying the estimates of the triangular inequalities for the q-norm, it suffices to find p such that

$$\|\vec{D}(f) - \vec{\Gamma_p}(f)\|_q \leq 2^{d/q}\lambda\varepsilon$$

$$\sum_{\iota \in \{-1,1\}^n} \left| D^{(\iota)}(f) - \Gamma_p^{(\iota)}(f) \right|^p \leq 2^n \lambda^q \varepsilon^q$$

$$\forall \iota \in \{-1,1\}^n, \left| D^{(\iota)}(f) - \Gamma_p^{(\iota)}(f) \right|^p \leq \lambda^q \varepsilon^q$$

$$\forall \iota \in \{-1,1\}^n, D^{(\iota)}(f) - \Gamma_p^{(\iota)}(f) \leq \lambda\varepsilon$$

Now we get that for all ι, p has to be greater to $p^{(\iota)}$ according to Theorem 12. In other terms, by defining $p_0 := \max_\iota p^{(\iota)}$ it ensures that $d_2 \leq \lambda\varepsilon$ and therefore we have $\delta_{p,q} \leq (1-\lambda)\varepsilon + \lambda\varepsilon$ which finishes the proof. □

Remark that the most important part of the approximation, and the smallest one we need to get, is the one regarding p. The one regarding q is numerically more stable. Moreover, q is just a classical 2^n dimensional vector norm which will be used for n being practically 2 (for images) of 3 (for volumes) and will not create troubles during the applications. Therefore we might at first suggest to get λ as close to 1 as possible which is not really clear according to Table. 4.2.

Another remark comes from the computation of q_0 where the value of the discrepancy norm of f is actually needed. In this case one can question themself whether all these calculations make sense if we anyway have to compute the discrepancy norm to get an estimator of it. However in the practical examples concerned with autocorrelation, we will compute this value only once for the complete registration procedure.

Example of lower bounds Here we introduce three toy examples of functions defined on $\mathbb{R}^2$, which we will use as test samples.

$$f_1 : [1, 50] \times [1, 50] \to \mathbb{R}$$
$$\mathbf{x} = (x, y) \mapsto \mathbf{1}_{[10,20[}(\mathbf{x}) + \mathbf{1}_{[30,40[}(\mathbf{x}) \tag{4.42}$$
$$f_2 : [1, 50] \times [1, 50] \to \mathbb{R}$$
$$\mathbf{x} = (x, y) \mapsto \sin(\mathbf{u}^T \mathbf{x}) \tag{4.43}$$
$$f_3 : [1, 50] \times [1, 50] \to \mathbb{R}$$
$$\mathbf{x} = (x, y) \mapsto \frac{1}{\|\mathbf{x}\|_2} \sin(\|\mathbf{x}\|_2) \tag{4.44}$$

In the definition of f_2 the vector $\mathbf{u}$ is characterized by two parameters: an angle α and a period

T, so that we can also write $f_2(x,y) = \sin\left(\frac{\cos(\alpha)x+\sin(\alpha)y}{T}\right)$

These examples are depicted on Fig. 4.6. It must be noted that the second and third images (or functions) are oscillating around 0 while the first one is non negative. This has a particular effect as we will see in the results of Table. 4.2.

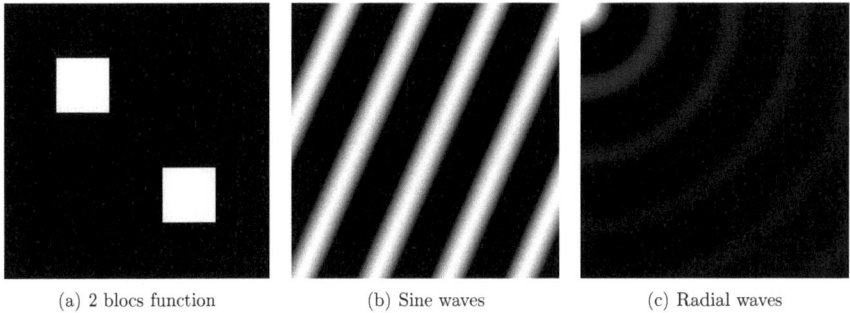

(a) 2 blocs function (b) Sine waves (c) Radial waves

Figure 4.6: Our dataset of 2 dimensional toy functions used in our paper.

Table. 4.2 gathers some results of the convergence and quality of the approximation for the different test functions with different parameters. It is divided into three groups of columns: the first one describes the different parameters used for the experiments (function, the maximal error expected ε, the percentage of ε needed for the thresholding process in the computation of p_0 $\alpha \in [0,1]$ and the repartition of the error between p-mean and q-mean approximations $\lambda \in [0,1]$), the results obtained for these parameter values with our approximations and estimations (p_0 and q_0 computed according to Corollary 4, the value of the approximation Γ_{p_0,q_0} and the value of the discrepancy norm) and the optimal exponents $\overline{p_0}$, $\overline{q_0}$ achievable with the given set of parameters. They are computed as follows:

$$\overline{q_0} = \min\left\{q \in \mathbb{N} : \|f\|_I^{(2)} - M_q(\overrightarrow{D}(f)) < (1-\lambda)\varepsilon\right\} \tag{4.45}$$

$$\overline{p_0} = \min\left\{p \in \mathbb{N} : M_{\overline{q_0}}(\overrightarrow{D}(f)) - M_{\overline{q_0}}(\overrightarrow{\Gamma_p}(f)) < \lambda\varepsilon\right\} \tag{4.46}$$

It is important to us that the estimated p_0 and q_0 stay close to the optimal values $\overline{p_0}$ and $\overline{q_0}$ so that the quality of the approximation is preserved, without getting into numerical overflows.

Concerning the first example of the two-bloc function, we see that our estimates are far above the optimal ones. This is due to the fact that the function is non-negative. In this case, $\forall q \in \mathbb{R}, M_q(\overrightarrow{D}(f)) = \|\overrightarrow{D}(f)\|_\infty = \min(\overrightarrow{D}(f)) = \|f\|_I^{(2)}$ so that no error is done on any non-negative functions during the q-means approximation.

The second function is an example of a (at least locally, due to the finiteness of the image) periodic function. In this case we notice that our estimations are close to the optimal ones (sometimes even smaller due to the discretisation of the computation of $\overline{q_0}$ and $\overline{p_0}$). Why this happens to be for such periodic signals has to be studied in more details and goes beyond the scope of this research.

In the last example of spherical waves, we can see that estimated powers are within two to three times bigger than the optimal ones yielding an error about half the maximal error expected.

What comes out of this table is that the choice of a good λ is vital to avoid numerical problems. Indeed, if not well tuned, it leads to high p or q powers and yields numerical overflows. Fig. 4.7 shows the impact of this λ on the estimated p_0 and q_0.

Table 4.2: Examples of approximation and lower bound estimations on some toy functions.

Function	ε	α	λ	p_0	q_0	Γ_{p_0,q_0}	$\|\cdot\|_I^{(2)}$	$\overline{p_0}$	$\overline{q_0}$
Blocs	1	0.75	0.75	29.88	1108.3	199.87	200	6	1
	1	0.75	0.5	46.92	553.82	199.92	200	6	1
	1	0.75	0.25	98.30	368.98	199.96	200	6	1
	1	0.5	0.5	24.24	553.82	199.84	200	6	1
	10	0.5	0.5	1.71	54.76	197.73	200	1	1
Sines	1	0.25	0.5	40.52	154.09	54.97	55.83	29	155
	1	0.3	0.5	39.32	154.09	54.96	55.83	29	155
	1	0.5	0.5	53.70	154.09	55.05	55.83	29	155
	1	0.75	0.5	90.59	154.09	55.16	55.83	29	155
	2.5	0.3	0.5	13.56	61.22	53.63	55.83	11	60
Radial	0.25	0.5	0.5	246.52	179.30	16.10	16.23	125	90
	0.5	0.5	0.5	121.39	89.30	15.98	16.23	62	42
	1	0.5	0.5	58.92	44.30	15.73	16.23	31	17
	1	0.5	0.4	74.52	36.80	15.75	16.23	38	13

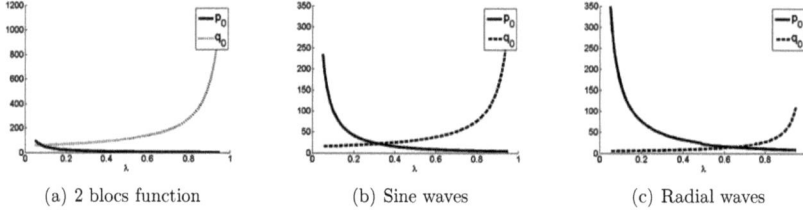

(a) 2 blocs function   (b) Sine waves   (c) Radial waves

Figure 4.7: Evolution of the values of p_0 and q_0 depending on the choice of λ.

It seems that, apart from the blocs samples, a λ within the range $[0.3, 0.7]$ gives a good compromise between a high q or a high p.

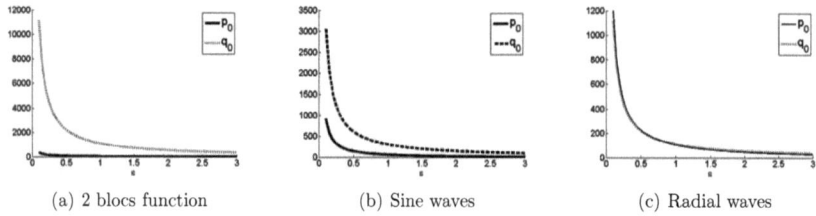

(a) 2 blocs function   (b) Sine waves   (c) Radial waves

Figure 4.8: Evolution of the values of p_0 and q_0 depending on the choice of ε.

4.2.2 Derivation of the correlation function

As in the one dimensional case, we introduce the discrepancy correlation function which we wish to minimize $J^{(n)}(t) := \|\tau_t f - g\|_I^{(n)}$, where f and g denotes two patterns we want to align.

Theorem 13 (Gradient computation of the discrepancy correlation for multivariate function). *The*

discrepancy correlation function's derivative $J^{(n)}(t)$ can be approximated in the following way:

$$\frac{\partial J^{(n)}}{\partial t}(t) \approx \frac{1}{2^{n/q}} \frac{\|\overrightarrow{\Gamma_p}(f_t-g)\|_q}{\|\overrightarrow{\Gamma_p}(f_t-g)\|_q^q} \sum_{\iota \in \{-1,1\}^n} \Gamma_p^{(\iota)}(f_t-g)^{q-1} \left\langle \theta_i^{(\iota)}(\cdot - t), \delta_{\chi^{(\iota)}} \right\rangle \qquad (4.47)$$

where we use the following notations:

$$\theta_i^{(\iota)}(x-t) = \int_{\iota\cdot[x[} \frac{\partial f}{\partial t_i}(s-t)\mathrm{d}\mu(s) \qquad (4.48)$$

$$\delta_{\chi^{(\iota)}} = \left(\frac{\chi^{(\iota)}(g-f_t)}{\|\chi^{(\iota)}(g-f_t)\|_p}\right)^p - \left(\frac{\chi^{(\iota)}(f_t-g)}{\|\chi^{(\iota)}(f_t-g)\|_p}\right)^p \qquad (4.49)$$

Proof. Let $\Delta_{p,q}[f,g](t) := \Gamma_{p,q}^{(n)}(f_t-g) = \dfrac{\left(\sum_\iota \Gamma_p^{(\iota)}(f_t-g)^q\right)^{1/q}}{2^n}$. We approximate $J^{(n)}(t) \approx \Delta_{p,q}[f,g](t)$.

Let $i \in \{1, \cdots, d\}$ we have

$$\frac{\partial \Delta_{p,q}[f,g]}{\partial t_i}(t) = \frac{1}{2^{n/q}} \frac{1}{q} \left(\sum_\iota \Gamma_p^{(\iota)}(f_t-g)^q\right)^{1/q-1} \frac{\partial \sum_\iota \Gamma_p^{(\iota)}(f_t-g)^q}{\partial t_i} \qquad (4.50)$$

$$= \frac{1}{2^{n/q}} \frac{1}{q} \frac{\|\overrightarrow{\Gamma_p}(f_t-g)\|_q}{\|\overrightarrow{\Gamma_p}(f_t-g)\|_q^q} \frac{\partial \sum_\iota \Gamma_p^{(\iota)}(f_t-g)^q}{\partial t_i} \qquad (4.51)$$

$$\frac{\partial \sum_\iota \Gamma_p^{(\iota)}(f_t-g)^q}{\partial t_i} = q \sum_\iota \Gamma_p^{(\iota)}(\tau_t f-g)^{q-1} \frac{\partial \Gamma_p^{(\iota)}(\tau_t f-g)}{\partial t_i} \qquad (4.52)$$

$$\frac{\partial \Gamma_p^{(\iota)}(f_t-g)}{\partial t_i} = \left\langle \theta_i^{(\iota)}(\cdot - t), \delta_{\chi^{(\iota)}} \right\rangle \qquad (4.53)$$

So that combining every components together we get

$$\frac{\partial \Delta_{p,q}[f,g]}{\partial t_i}(t) = \frac{1}{2^{n/q}} \frac{\|\overrightarrow{\Gamma_p}(f_t-g)\|_q}{\|\overrightarrow{\Gamma_p}(f_t-g)\|_q^q} \sum_{\iota \in \{-1,1\}^n} \Gamma_p^{(\iota)}(f_t-g)^{q-1} \left\langle \theta_i^{(\iota)}(\cdot - t), \delta_{\chi^{(\iota)}} \right\rangle \qquad (4.54)$$

□

4.3 Practical considerations

In this section we want to give some practical details regarding the implementation of the derivative. Indeed, Eq. 4.47 seems quite complicated and not really suitable for practical use.

First let's decompose all the components of the formula. The first one $\frac{1}{2^{n/q}} \frac{\|\overrightarrow{\Gamma_p}(f_t-g)\|_q}{\|\overrightarrow{\Gamma_p}(f_t-g)\|_q^q}$ is just a multiplicative factor and can be simplified as $2^{-n}\Gamma_{p,q}(f_t-g)^{1-q}$ which depends neither on i nor ι.

The sum on the right can be understood as a scalar product in $\mathbb{R}^{2^n}$. As ι travels all over $\{-1,1\}^n$, each $\Gamma_p(\iota)$ takes value in $\mathbb{R}$ (and actually only positive values as already seen). So does the scalar product (in $\mathbb{R}^n$ this time) $\left\langle \theta_i^{(\iota)}(\cdot - t), \delta_{\chi^{(\iota)}} \right\rangle$; so that altogether the sum can be interpreted as $\left(\overrightarrow{\Gamma_p(f_t - g)}^{q-1}, \overrightarrow{\Theta_i(t)}\right)$ where

$$\overrightarrow{\Theta_i}(t) = \left[\int_{x \in \mathbb{R}^n} \theta_i^{(\iota)}(x-t)\delta_{\chi^{(\iota)}}(x)\mu(x)\right]_{\iota \in \{-1,1\}^n}$$

. The power of vector $\overrightarrow{\Gamma_p}$ is to be understood componentwise.

In many practical cases, the μ measure will be slowly varying (if not completely uniform on a compact support) and discrete. In this configuration, and in particular in the case of image, $n = 2$, we have (assuming $\iota = (1,1)$ i.e., the summation start from the $(-\infty, -\infty)$ corner; the other cases are computed similarly)

$$\theta_i^{(\iota)}(x-t) = \int_{\iota \cdot [x[} \frac{\partial f}{\partial t_i}(s-t)\mathrm{d}\mu(s) \tag{4.55}$$

$$" = " \sum_{l=-\infty}^{x_1} \sum_{m=-\infty}^{x_2} \frac{\partial f}{\partial t_i}(l-t_1, m-t_2)\mu(l,m) \tag{4.56}$$

The partial derivative can be approximated by finite differences; in the continuous case, the fundamental theorem of calculus applies.

$$\theta_i^{(\iota)}(x-t) = \sum_{l=-\infty}^{x_1} \sum_{m=-\infty}^{x_2} \left(f(l-t_1+1, m-t_2) - f(l-t_1, m-t_2)\right)\mu(l,m) \tag{4.57}$$

$$\approx \sum_{m=-\infty}^{x_2} f(x_1+1-t_1, m-t_2)\mu(x_1+1, m) \tag{4.58}$$

and the last formula tells that we only need to do some kind of integral images in $n-1$ direction (apart from the one in which we wish to differentiate). This is easy and efficient to compute when working on images.

However even if the theory is still interesting and well-founded, some efforts still have to be done to avoid numerical overflow when dealing with rather big images (over 200 pixels wide)

Part III

Other approaches towards structures and irregularities in images

Chapter 5

Distance transform methods in image processing

In this chapter we give some ideas for structures and irregularities comparison by means of distance transform algorithms. We first give a review of the distance transform approaches as they are used in image processing and then give some details about local dissimilarity analysis. In a last section we introduce a novel approach for grey-level distance transform, and see its applicability in some image comparison tasks.

5.1 Distance transforms for image processing

This section is intended as a short review of already existing methods in the context of image comparison. We start by recalling the problems and solutions for considering distance transforms on binary images. We then introduce some ideas for extension to grey-level images. In all cases similarity measures are derived from the distance transform representations.

5.1.1 Distance transform on binary images

5.1.1.1 Motivation: comparison of edge detection algorithms

The motivating idea as introduced by Baddeley in 1992 [Bad92] was to assess the quality of different edge detection algorithms. In his work the author shows that traditional quality measures, whether statistical or localisation measures, show some weakness regarding quality assessment.

In particular, he considers the example of the Figure Of Merit [AP79] (FOM) defined for an image f and an estimation g (both binarized) as

$$FOM(f,g) = \frac{1}{\max\{|A_f|,|A_g|\}} \sum_{x \in A_g} \frac{1}{1+\alpha d(x,A_f)^2} \qquad (5.1)$$

where $A_f = \{x : f(x) = 1\}$, and $A_g = \{x : g(x) = 1\}$ (note that we independently use $x \in A_f$ or $x \in f$ in an abusive way) and $|\cdot|$ is the cardinality of the set, α is just a weighting constant. This measure is not symmetric and takes into account only false positive errors and not false negatives (some authors prefer to say Type I and Type II errors). It means that a certain pixel x will be considered as taking part to the FOM score if and only if it is part of A_g (the estimation) but not in A_f (the model). So that the pixels omitted by the estimation are not given any importance.

This measure is also independent of the arrangements of the errors and some critical examples can be found where two drastically different observations compared to a same reference one will have the same FOM score. Fig. 5.1 (taken from [PM82]) illustrates this idea. When compared to the central pattern, both left and right patterns have the same FOM score even though they look completely different.

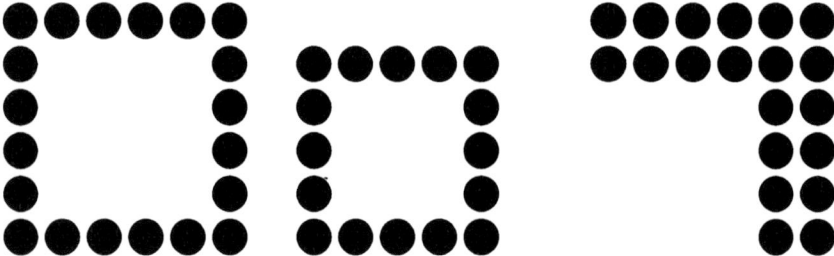

Figure 5.1: Examples of two completely different binary images which have the same FOM score when compared to a same third one.

This is something we should try to avoid when designing a score for image quality or comparison.

5.1.1.2 Binary images as sets

The basic idea for using distance transform based algorithms relies in representing binary images as set. Given a binary signal f defined on $\mathbb{R}^n$, it can actually be represented as a set $A_f = \{x \in \mathbb{R}^n : f(x) = 1\}$. This is the key idea in using distance transforms in binary image processing and analysis. As we see, the Hausdorff measure acts as a robust distance between sets and can be computed, in a binary case, using distance transforms (see Eq. 5.6 for more details).

5.1.1.3 The distance transform

The distance transform works by affecting at each point or pixel of the domain of definition, the distance to the closest point of the background (modeled by a binary 0). So for a proper definition, we need to define a distance transform on a metric space with distance d as follows:

Definition 15 (Distance transform). *Let f be a binary function (i.e. with values in $\{0,1\}$) on the metric space $(\mathbb{R}^n, d)$. Its distance transform (to the background) is defined as*

$$\forall x \in \mathbb{R}^n, \ DT_f(x) = \min_{y:f(y)=0} d(x,y) \qquad (5.2)$$

Note that this distance map yields non vanishing values only for foreground points (modeled with a binary 1).

5.1. DISTANCE TRANSFORMS FOR IMAGE PROCESSING

Moreover, we can define a dual definition based on the distance transform to the foreground. In that case, only points from the background will get non vanishing values. And the distance transform to the foreground should be prefered for edge comparison, due to the local property of edges.

In practical applications we deal with discrete metrics, and people are for instance using:

- Taxicab metric or Manhattan distance. It corresponds to the L^1 norm; or in terms of neighborhood, to a 4-neighborhood scheme

- Chessboard distance or Chebyshev distance. It corresponds to the L^∞ norm; it is associated to a 8-neighborhood scheme

- Euclidean norm. It corresponds to the L^2 case or the classical aerial distance.

However, any metric can be used here, and other Minkowski distances might also be some interesting choices. As we will see later, we can also approximate the Euclidean distance a smart way to make it tractable on practical examples.

Applications of distance transforms Besides some application in pattern recognition and image analysis, the distance transforms have been widely used for different tasks.

- Skeletonisation [Cha07] is the process of marking an equidistant line to the boundaries of an object. It can be used for instance in pose estimation [TDSL00].

- Solution of the Eikonal equation [VV90]. This example of non-linear partial differential equation describes the propagation of some approximations of waves.

- Pathfinding. This application helps a rover finding its way in an unknown area [CWT+01]

- Motion planing [RN02]. This application is similar to the previous one but can also be used for instance in artificial intelligence in video games.

5.1.1.4 Implementation details

There are mainly two ways of implementing the distance transform algorithms on a discrete pixel grid: sequential or parallel. In the parallel approach, the distances are propagated iteratively according to a certain mask $(c_{ij})_{ij=-s}^{s}$ (with s being the size of the mask) representing some kind of cost of displacement. That means that it is assumed that pixel (i,j) and $(0,0)$ are separated by a distance c_{ij}. Formally speaking, the distance transform at a location x is the minimum over all so far propagated distances, at iteration m:

$$DT_f(x_1, x_2)^{(m)} = \min_{i,j} DT_f(x_1 + i, x_2 + j)^{(m-1)} + c_{i,j} \tag{5.3}$$

and the process is initialised with infinity values on foreground pixels. These iterations occur until no more changes appear. We should note that the mask has to fulfil some properties regarding comparison of its coefficient, to ensure the metric axioms of the results (mainly because of the triangle inequality).

On the other side, a sequential algorithm can be implemented by a simple two pass approach where the original mask is split into two parts, one being passed from the top-left corner of the image to the bottom right one, and the other half in the opposite direction.

Borgefors [Bor86] described an approach to optimise the coefficients of the mask in order to ensure a result close enough to the Euclidean norm without needing much computations. It comes out that the algorithm can be implemented in linear time with less than 2% difference to the exact Euclidean distance transform.

5.1.1.5 Hausdorff metric

Definition 16 (Hausdorff distance). *Given two sets $A, B \subset \mathbb{R}^n$, the Hausdorff distance is defined as*

$$H(A,B) = \max(\sup_{x \in A}(d(x,B)), \sup_{y \in B}(d(y,A))) \tag{5.4}$$

It should be noted that this definition is actually a symmetrized version of the relative Hausdorff measures $\sup_{x \in A}(d(x,B))$ and $\sup_{y \in B}(d(y,A))$. The symmetrical definition is the only one we use in the sequel and therefore we refer to it as the Hausdorff metric.

Interestingly, it is one of the only measures (see Subsection 3.3.2 for more details) showing a monotonicity property with respect to translation of one of the sets [BMNMR08].

Property 5. *The autocorrelation function based on the Hausdorff distance is monotonic with respect to shifting of a pattern. Even more, we have for a translation vector $t \in \mathbb{R}^n$ and a binary function f that*

$$H(f, \tau_t f) = \|t\| \tag{5.5}$$

Proof. The proof has four components: showing both inequalities for both directed Hausdorff distances. However, if we write $g = \tau_t f$ then we have $f = \tau_{-t} g$ and it is equivalent to show that $\sup_{x \in f}(d(x, \tau_t f)) = \|t\|$

Now consider a point $x \in f$, it holds

$$\min_{y \in \tau_t f} d(x,y) \le d(x, \tau_t x)$$

$$\le \|t\|$$

$$\max_{x \in f} \min_{y \in \tau_t f} d(x,y) \le \|t\|$$

Now to prove the other direction, we assume that $\max_{x \in f} \min_{y \in \tau_t f} d(x,y) = \epsilon < \|t\|$. It means that $\forall x \in f, d(x, \tau_t f) \le \epsilon$ which is a nonsense. □

Baddeley [Bad92] proved the following result:

$$H(A,B) = \sup_{x \in X} |d(x,A) - d(x,B)| \tag{5.6}$$

where X corresponds to the bounded domain on which we are working.

Proof. The original proof can be found in [Bad92]. First notice that for x being in both the pixel raster and in the (set representing the) first function, we have that $d(x,A) = 0$ and therefore $|d(x,A) - d(x,B)| = d(x,B)$ and we get a similar result for x in the second set. Therefore, $|d(x,A) - d(x,B)| \ge H(A,B)$ holds.

5.1. DISTANCE TRANSFORMS FOR IMAGE PROCESSING

Now we have a look at the opposite inequality. Due to the triangle inequality of the underlying distance, it holds that $d(x, A) \leq d(x, y) + d(y, A)$. Now for a fix x in the pixel raster, we have, for all strictly positive ε that it exists a point y in B such that $d(x, y) < d(x, B) + \varepsilon$, which, combined with the previous triangular inequality, yields $d(x, A) < d(x, y) + d(y, A) < d(x, B) + d(y, A) + \varepsilon$ which implies $d(x, A) - d(x, B) < d(y, A) + \varepsilon$. Now we can intertwine the roles played by A and B and take the suprimum, which leads to the following $\sup_{x \in X} |d(x, A) - d(x, B)| < H(A, B) + \varepsilon$. ε being chosen as wished finishes the proof. □

5.1.1.6 Image comparaison with distance transforms

As seen in the previous section, binary image comparison can be done based on the distance transform and Hausdorff measure. However, due to the use of a max in it, it is highly biased by the presence of noise and if two shapes differ only by a single spike on the border, this spike will be responsible for the whole Hausdorff measure, without noting that the two objects actually do look similar.

To overcome this problem, Baddeley proposed to replace the max by a p-norm approximation, and to average all contributions.

Definition 17 (Baddeley's similarity between binary images). *Two binary images f and g defined on a finite (discrete) domain $D \subset \mathbb{R}^n$ can be compared using the following coefficient:*

$$\Delta_b^p(f, g) := \frac{1}{|D|} \left(\sum_{x \in D} |d(x, f) - d(x, g)|^p \right)^{1/p} \tag{5.7}$$

This approach tends to average out the local irregularities. Indeed, assume two images are the same up to a few points. Then the value inside the sum will be always 0 but on the irregularities. Therefore, as long as they are not many such errors or unwanted phenomenon, the global dissimilarity value will be kept low.

Note however that this is not the only possibility to overcome this noise problem. One can for instance consider taking the Kth biggest distance instead of the max, or a weighted version of all of them. More details about some modified Hausdorff distances are given in [ZSD05]

5.1.2 Extensions to grey level images

How one should extend this distance transform process to gray-level images is not clear. We will here review some ideas proposed in the past.

5.1.2.1 Distance transforms along a path

Gray weighted distance transform (GWDT) The Gray Weighted Distance between two pixels is defined in [LM70] and [VV90] as the smallest weighted sum of grey level values along the discrete path between these two points. It corresponds to the surface area estimation under a curve path. It relies on the following cost between two adjacent pixels :

$$w_{GWDi} = \frac{1}{2}(f(t_i) - f(t_{i+1})) \times ||t_i - t_{i+1}||, \tag{5.8}$$

where $||t_i - t_{i+1}||$ is the spatial distance between the two pixels and $f(t_i)$ is the grey value of f for pixel t_i.

Weighted distance transform on curved space (WDTOCS) The path between two points is defined as an $n+1$ dimensional path constraint to lie on the hyper-surface defined by the grey level values. In [Toi96], it is expressed as the length of the shortest geodesic path between these two points. It relies on the following cost :

$$w_{WDOCSi} = \sqrt{(f(t_i) - f(t_{i+1}))^2 + ||t_i - t_{i+1}||^2}. \tag{5.9}$$

The main problem with this distance is the inconsistency of the units. By considering grey level values as $n+1$ image dimensions, the method mixes spatial and intensity values. To cope with this problem, images values can be scaled by a coefficient, masking the inconsistency.

Continuous distance transform (CDT) This distance transform is based on a generalisation of the "white pixel" and "find the nearest white neighboring pixel", see [AP00]. The "white pixel" becomes the "maximum bright value" and the "find the white nearest neighboring pixel" is replaced by "accumulate a maximum bright value on the neighborhood".

This last definition seems more an "ad-hoc" response to the addressed problem without straight interpretation. We prefer the simpler definition of the GWDT and the WDTOCS.

Each of these distance transforms but the CDT can be computed in a fast way with a good approximation by a two-pass algorithm, see [IT05].

5.1.2.2 Wilson's approach

This approach, as well as the 2 following ones, was introduced in order to extend the concept of distance transform to sampled images. We see that once again, we can define a similar dissimilarity measure for image comparison purposes.

The approach introduced by Wilson et al. [WBO97] aims at extending the previous algorithm to one capable of handling grey-level images. The idea is to embed an n-dimensional signal into a n+1-dimensional space and to compute distances from each point of this space to the subgraph of the signal.

Definition 18 (Subgraph of a signal). *Given a signal f defined on $X \subset \mathbb{R}^n$ whose values are in a given set Y (i.e. $f : X \to Y$), its subgraph Γ_f is defined as*

$$\Gamma_f = \{(x, y) \in X \times Y : y \leq f(x)\}$$

Remark that we could generalise to any higher dimensional output sets, as long as they are totally ordered.

Note that n will typically have value 1 (for proper signals) or 2 (for images). Y is the set of possible grey values of the signal; for instance, for an 8-bit image we have $Y = \{0, \cdots, 255\}$.

From now on, if needed we will denote by *spatial coordinates* any point in X and by *intensity* or *illumination* values in Y.

5.1. DISTANCE TRANSFORMS FOR IMAGE PROCESSING

Now, let D be the containing space, i.e. $D = X \times Y$, we compute, for each point $p = (x, y) \in D$ the distance to the subgraph of f:

$$d(p, \Gamma_f) = \inf_{p' \in \Gamma_f} d(p, p') \qquad (5.10)$$

Remark that expression 5.10 is general and one needs to specify the distance used inside the inf. Wilson et al. have proposed a distance that separately takes into account spatial coordinates and intensity values.

Definition 19 (Wilson's distance). *We can define a distance on $X \times Y$ as*

$$\forall p = (x, y),\ p' = (x', y') \in D,\ d_W(p, p') = \max(d(x, x'), |y - y'|)$$

where $d(x, x')$ defines, for instance, the Euclidean distance between the spatial coordinates.

We can now introduce the notion of upper level set, which is a useful concept for the derivation of a metric which makes use of distance transform just like in the binary case.

Definition 20 (Upper-level set). *Given a signal f, we define an upper-level set at y as*

$$X_y(f) = \{x \in X : f(x) \geq y\}$$

Finally plugging definitions 19 and 20 into equation 5.10 yields Wilson's grey level distance transform:

$$\forall p = (x, y) \in D,\ DT_W[f](p) = d(p, \Gamma_f) = \inf_{z \in Y} \{\max(d(x, X_z(f)), |y - z|)\} \qquad (5.11)$$

As a consequence only a finite number of binary distance transforms have to be computed to get Wilson's grey level distance transform.

However, we propose another way to write this down, as it is rather unclear how to understand, in a physical sense, the distance used here. Indeed, if we consider a one-dimensional signal, then the set X is a set of given time stamp while Y represents quantitative values (price of an action for instance) and one should be careful when taking a max on variables with different signification. Therefore we propose to add a normalization parameter which should be there to make sure both spatial and intensity variables are comparable. This yield the following normalized distance:

$$d_W^\rho(p, p') = \max(d(x, x'), \rho|y - y'|) \qquad (5.12)$$

Embedding a signal into a higher dimensional space has another drawback. It makes the visualization of the distances hard. Indeed, considering a 2-Dimensional signal (an image) the distance transform using the approach of Wilson et al. yields a 3-Dimensional map which is no longer easily displayed.

A last idea suggested is to cut off values higher than a threshold c in the definition of the distance transform.

Definition 21 (Thresholded distance). *Let c be a positive real number. The following expression can be used as a distance for the distance transform:*

$$d_W^{\rho\,*}(p, p') = \min\{d_W^\rho(p, p'), c\}$$

This formulation allows to control the highest value achievable in the distance transform and to accelerate the computation. Indeed, by considering a thresholding of the distance, we can directly get rid of all the distance transforms that appear at an intensity distance of more than the thresholding (more details are given in [WBO97]). Moreover, as we see in the images depicted on Fig. 5.2 the c parameter allows to show some fineness in the details. For small values (top rows) all the details even very small appear and the distance transform map is actually rather poor (in terms of information contained), as the this c value acts also as a saturation parameter. However with c getting to infinity (bottom rows), only bigger structures remain in the distance transform.

From left to right, only the normalisation parameter is changing. This allows to give more impact to the intensity changes or to the spatial differences in Wilson's distance. A higher normalisation tends to emphasise the intensity differences and thereafter makes the distance transform reach the saturation level faster. As we would expect a ρ in the order of magnitude of the signal's dynamic seems to be a good choice.

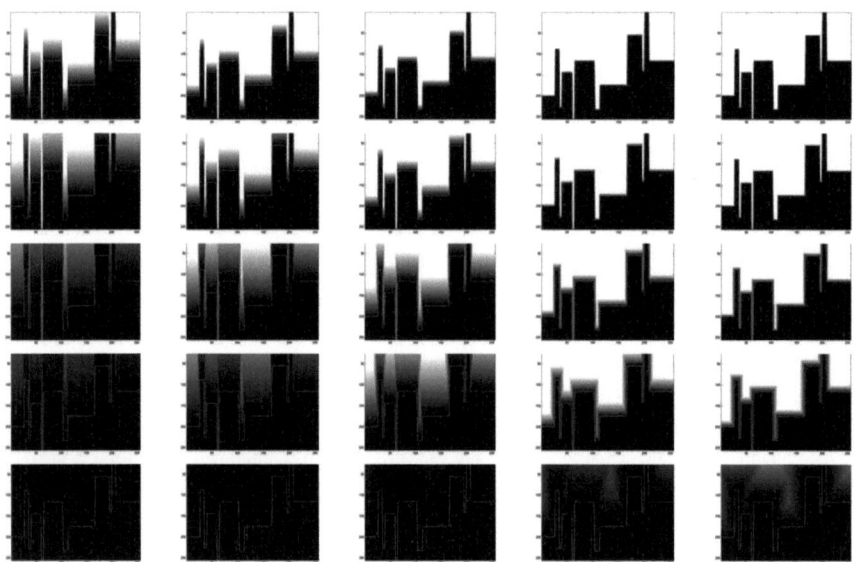

Figure 5.2: Influence of the different saturation and normalization parameters of Wilson's Distance Transform. Each row fixes a c (1,2,5,10,100, from top to bottom) while each column fixes a ρ parameter (0.5,1,2,8,16 from left to right). The upper left corner shows the signal used for the transformation. The reference signal is actually of size 256 and ranges from about -2 to 8.

5.1.2.3 Molchanov's approach

Another extension to higher dimensional space has been proposed by Molchanov and Terán [MT03]. In their work the authors introduce a distance transform which is also based on upper-level sets and sum them together with some predefined weight. They no longer consider the illumination coordinates (except intrinsically for the definition of the upper level sets). This simplifies the choice for a

5.1. DISTANCE TRANSFORMS FOR IMAGE PROCESSING

normalization factor, in opposition to Wilson's approach.

Definition 22 (Molchanov's Distance Transform). *Given a real-valued signal f defined on a domain $X \subset (\mathbb{R}^n, d)$, with distance metric d, we can define a real-valued distance transform as follows:*

$$DT_M[f](x) = \sum_{y \in Y} \mu_y d(x, X_y(f))$$

where d defines either a metric on the space $\mathbb{R}^n$ (for instance the Euclidean distance on the spatial coordinates), or, in an abusive way, the distance between a point and a subset $X_y(f) \subset \mathbb{R}^n$ and $\{\mu_y\}_{y \in Y}$ denotes some weighting factors.

The interesting point with this definition is that it does not create any artifact for the visualization by adding a dimension to the signal. Moreover it has been well developed in a mathematical sense and allows one to invert this procedure (at least for discrete finite signals; which is the case for images); it means that one can compute a distance transform map from an image or recover an image from a distance transform map.

Figs. 5.3 show the impact of discretising the intensity space over different values. It appears that the structure of the image with higher number of intensity levels is very similar to the one with low number of intensity levels.

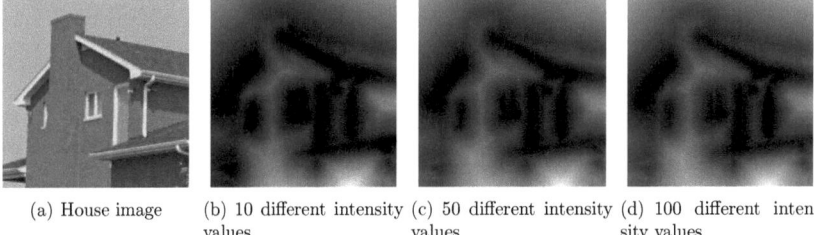

(a) House image   (b) 10 different intensity values   (c) 50 different intensity values   (d) 100 different intensity values

Figure 5.3: Molchanov distance transform on the House image. First row show the results with varying number of discretisation steps of the intensities while second row show varying value of thresolding parameter.

As before, we can introduce a thresholded variant of the given distance: every distance higher then a given threshold c are fixed to c. This allows a better control on the range of values achievable by the distance transform. It also allows to consider dissimilarity within a circle of a given radius around each pixel. Indeed, as it considers only distances which are beneath a certain the threshold c, it cannot give any details for structures appearing at a distance further than c from the current pixel.

Figs. 5.4 show the impact of choosing different thresholding values. It appears that the details appearing with the strongest thresholding ($c = 10$) are finer then the one appearing with higher thresholding values (see for instance Fig. 5.4(d) and 5.4(h), which have almost infinite dissimilarity tolerance).

5.1.2.4 Coquin's approach

A last extension we would like to present here is the work of Coquin and Bolon [CB01]. Their work is based on a similar idea as the one of Baddeley. The idea here is to embed the grey intensity value as an

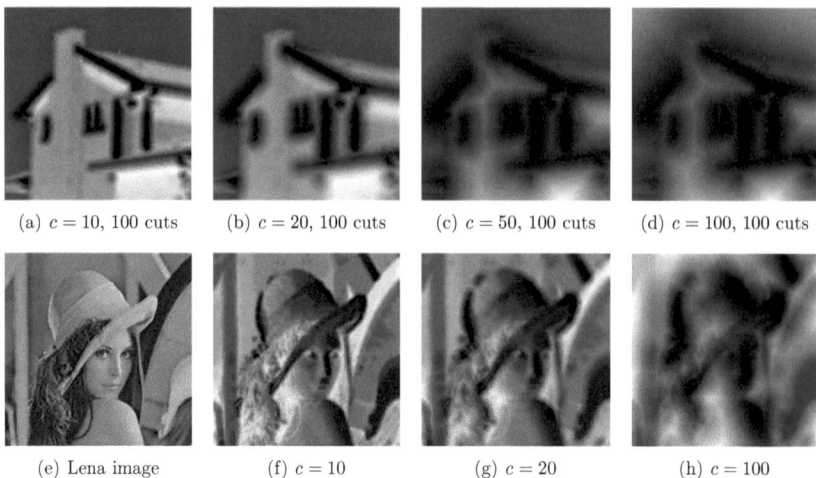

Figure 5.4: Molchanov distance transform with different thresholdings of the distances.

altitude and consider the image as a binary set in 3 dimensions. Formally, an image $f : \mathbb{R}^2 \to \mathbb{R}$ can be represented by a 3D surface $A_f = \{x = (x_1, x_2, x_2) \in \mathbb{R}^3 : f(x_1, x_2) = x_3\}$. Using this representation, the distance transform can directly be generalized to higher dimensional images and more complex images.

The interesting point with this representation relies on the possibility to approximate the Euclidean distance transform by just two-fold filtering the image [Bor84]. Besides these practical considerations, this approach takes into account the whole space in the representation instead of limiting the comparisons to the points above the subgraph, as it was the case in the two previous extensions. As a consequence of that, the derived similarity measure is invariant with respect to video inversion.

5.1.2.5 Comparing grey-level distance transforms

While we now have different ways to extend the distance transform to grey-level (or sampled) signals we have similar ways of comparing two such representations as in the work of Baddeley [Bad92]. This extension of the error metric to Wilson's version has already been used but to our knowledge no such image comparison metrics have been introduced considering Molchanov's distance transform. In the sequel, we denote by Δ_W^p the dissimilarity metric based on Wilson's approach based on Baddeley's binary dissimilarity measure *i.e.*

$$\Delta_W^p(f, g) := \frac{1}{|D|} \sum_{x \in D} |DT_W^\rho[f](x) - DT_W^\rho[g](x)|^p \tag{5.13}$$

Similarly we introduce Molchanov's comparison coefficient

$$\Delta_M^p(f, g) := \frac{1}{|X|} \sum_{x \in X} |DT_M^\mu[f](x) - DT_M^\mu[g](x)| \tag{5.14}$$

It should be noted that the averaging is done on two different domains, due to the fact that

5.2 Local dissimilarity maps

In such approaches, the dissimilarities are computed on a local neighbourhood of each given pixels. We refer to the works of Baudrier *et al.* [BMNMR08] for the binary case and to Morain *et al.* [MNLR09] for the case of sampled functions.

The idea is to constraint the Hausdorff distance to be taken only in a neighbourhood within a certain distance of the considered pixel. This needs however to specify the size of a neighborhood which can be problematic. Indeed when searching for dissimilar features, the size of the window should be fitted to a size close to the one of the features. If the features are too small (or the window too big) the local dissimilarities will not manage to catch their essence. Conversely, if the features are too big (or the window too small) only part of the feature will be considered for comparison leading to a lack of information. In their work Baudrier *et al.* introduced an adaptive version of the Hausdorff measure which automatically finds a locally optimal window size.

With this in mind, the authors suggest to make an explicit use of distance to the frontier or border of the window as follows:

$$H_w(f,g) = \max\{h_w(f,g), h_w(g,f)\}, \text{ with} \tag{5.15}$$

$$h_w(f,g) = \max_{x \in f \cap w} \left\{ \min\left(\min_{y \in g \cap w} d(x,y), \min_{z \in Fr(W)} d(x,z) \right) \right\}, \tag{5.16}$$

note however that this last expression, only if both $f \cap w$ and $g \cap w$ are non trivial and need some adaptation otherwise. In the last formula, $Fr(w)$ denotes the frontier of the window.

The interesting point with such measures is that it allows us to compute efficiently local dissimilarities using the following formula

Definition 23 (Local dissimilarity maps for binary images). *Given two images f and g, a Local dissimilarity map (LDMap) can be computed as*

$$LDMap(x) := |f(x) - g(x)| \max\{d(x,f), d(x,g)\} \tag{5.17}$$

This last definition can be extended to the case of grey-level images by replacing the distance transforms $d(x,f)$ and $d(x,g)$ by any of the variant proposed above and this leads to a characterisation of local dissimilarities between two images [MNLR09].

5.3 Multiscale analysis and distance transforms

In this section we want to introduce a novel method to generalise the distance transforms to grey-level images [BMN13]. Besides intensity information, we can interpret an image as being an overlap of different edges or structures at different scales. This actually reminds how the brain works by detecting structures at different refinement levels [CKP+07]. Therefore, in this approach, we make use of a scale-space representations in order to generate edges at different scales. These edges are then used as an input of a distance transform based algorithm.

To motivate this choice we recall the principal idea behind the introduction of distance transforms and image comparison with the Hausdorff measure (Bad92b): it was first introduced to assess the quality of edge detection algorithms. Therefore we go back to the idea of comparing edges of an image, but represented at different scales.

We give first some theoretical background about scale space representations and scale dependent edge detections of images before giving some details on how to apply it to image representations using distance transform algorithms.

5.3.1 Scale based edge detection

5.3.1.1 Basic scale-space theory

The axioms given to describe a general framework for scale space analysis of images are not always the same. For instance, [tHRF93] derives a linear scale space as a physical phenomena by defining an optimal aperture. By a direct dimension analysis (in a physical sense) the authors derive a general equation that the aperture should fulfil. Adding a translation invariance constraint, boundary conditions for a 0 and an infinite aperture, and a semigroup property (*i.e.* the combination of two scalings s_1 and s_2 is again a scaling of parameter s_3 depending only on s_1 and s_2), leads to the fact that the linear scale space can be achieved only by convolution with a exponential at a power p. The author then claims that the separability of the convolution kernel is needed, and that the physics fixes the power $p = 2$ as the only candidate. So in conclusion, generated with such axioms or physical prerequisites, the only appropriate candidate for linear scale spaces is the convolution of the Gaussian kernel where the radius plays the role of the scale (up to a square). These axioms are globally equivalent to the ones chosen by Koenderink [Koe84] or Lindenberg [Lin94].

However, this is not the only way to give an axiomatic description of linear scale space. Felsberg and Sommer [FS04] for instance decided to consider the scale space as the observation of an image from different distances.

Definition 24 (Observation). *Given an image f, considered as an n-dimensional signal with finite energy, we can define an observation transformation Φ which yields an observation of our image as:*

$$f(x_0; s) = \Phi(f, x_0, s) = \int_{\mathbb{R}^n} \phi(f(x), x_0, x, s) \mathrm{d}x \tag{5.18}$$

An example of such function is given by $\phi(f(x), x, x_0, s) = f(x) \exp(-\frac{\|x-x_0\|^2}{2s^2})$ which corresponds to a blurred observation of the image.

And a scale and rotation invariant scale-space can be derived from such observations.

Definition 25 (Scale and Rotation Invariant Scale-Space). *A scale and rotation invariant scale space is an observation transformation Φ which fulfils the following axioms:*

1. *Φ is linear:*

$$\forall \lambda \in \mathbb{R}, \ \Phi(\lambda f, x_0, s) = \lambda \Phi(f, x_0, s)$$

2. *Φ is shift-invariant:*

$$\Phi(\tau_t f, x_0, s) = \Phi(f, x_0 - t, s)$$

5.3. MULTISCALE ANALYSIS AND DISTANCE TRANSFORMS

3. Φ fulfills the semigroup property:

$$\exists S: \mathbb{R}^+ \times \mathbb{R}^+ \to \mathbb{R}^+, \quad \Phi(\Phi(f,\cdot,s_1), x_0, s_2) = \Phi(f, x_0, S(s_1, s_2))$$

4. Φ is scale and rotation invariant:

$$\exists T: \mathbb{R}^+ \times \mathbb{R}^+ \to \mathbb{R}^+, \quad \forall a \in \mathbb{R}^+, R \in SO(2), \quad \Phi(f(aR\cdot), x_0, s) = \Phi(f, aRx_0, T(s, a))$$

5. Φ preserves positivity:

$$\forall f > 0, \quad \Phi(f, x_0, s) > 0$$

Note that it appears that working with these sets of axioms allows us to derive more general scale space representations.

An even more general theory is about the α scale space. Their characteristics are well studied in [DFDGtHR04]. α scale spaces $\Phi: L^2(\mathbb{R}^n) \times \mathbb{R}^+ \to \mathbb{R}^n, \forall s \in \mathbb{R}^+, u(x; s) = \Phi[f, s](x)$ are assumed to be fullfilling the following axioms

1) f should be square integrable with compact support

2) The scale space is translation invariant: $\forall a \in \mathbb{R}^d, \Phi[\tau_a f; s] = \tau_a \Phi[f; s]$

3) Scale (almost) invariance: $\forall s, \exists s\prime, \Phi[D_{\delta^{-1}} f; s] = D_{\delta^{-1}} \Phi[f; s\prime]$, where the scaling operator that maps s to $s\prime$ is strictly increasing, continuous from $[0, \infty[$ to itself

4) Positivity: $f \geq 0 \Rightarrow \Phi[f; s] \geq 0$

5) The scale space should be linear ($\Phi[\lambda f + g; s] = \lambda \Phi[f; s] + \Phi[g; s]$) and bounded for all s from $L^2(\mathbb{R}^n)$ to a) $L^2(\mathbb{R}^n)$ or b) to $L^\infty(\mathbb{R}^n)$

6) Semigroup property: $\Phi[\Phi[f; s_1]; s_2] = \Phi[f; s\prime]$, where $s\prime$ can be considered to be the sum of both s_1 and s_2 (otherwise a new parametrisation is still possible)

7) The scale space should fulfil a (potentially weak) causality constraint i.e. all information at a scale $s_1 > s_2$ can be recovered for the finer information at scale s_2

8) It should converge, in L^2, to the original signal, for the scale getting closer to 0

9) For $\rho \in SO(n)$ we have a rotation invariance: $R_\rho(\Phi[f; s]) = \Phi[R_\rho f; s]$

10) The average intensity should be invariant: $\|\Phi[f; s]\|_{L^1} = \|f\|_{L^1}$, for $s > 0$ and $f \geq 0$

11) The entropy $\varepsilon(\Phi[f; \cdot])(s) = -\int_{\mathbb{R}^d} \Phi[f; s](x) \ln(\Phi[f; s](x)) \, dx$ is growing with $s \to \infty$

It can be shown (see [DFDGtHR04] for the details)

Property 6. *Under the above described axioms, we have*

$$\exists K_s \in L^1(\mathbb{R}^d) : \Phi[f; s] = K_s * f \tag{5.19}$$

The scale space is therefore represented as an integral operator with translation invariant kernel $K_s(x) := \overline{K_s}(\|x\|)$. Moreover it follows that $K_s \geq 0$ and that $\int_{\mathbb{R}^n} K_s(x) = 1$; it is also the inverse Fourier transform of $\omega \mapsto e^{-\|\omega\|^{2\alpha}s}$ with $0 < \alpha \leq 1$.

For $\alpha = 1$ we get the well-known Gaussian scale space while $\alpha = 1/2$ yields the Poisson-scale space, as used in the monogenic context (see [FS04]).

As a consequence, we can see that the α scale spaces arise from the following pseudo-differential boundary value problem:

$$\begin{cases} u_s = -(-\Delta)^\alpha u \\ \lim_{s \searrow 0} u(\cdot, s) = f \end{cases} \quad (5.20)$$

end we get that the case $\alpha = 1$ yields the heat equation and therefore the Gaussian scale space, while $\alpha = 1/2$ yields the Poisson problem with the Poisson scale space associated.

In our case, we stick to the Gaussian case. Finally, we can build a multiscale edge detection with such a scale space representation by considering zero-crossing of the second derivative. Interestingly, due to the convolution property, the scale space acts also as a regulariser when one wants to compute derivatives.

Indeed a function f which could potentially be nowhere differentiable has a differentiable scale space representation thanks to the convolution with the Gaussian kernel, and we have:

$$\partial_x \Phi[f; s](x) = \partial_x (K_s * f)(x) = ((\partial_x K_s) * f)(x) = (K_s * (\partial_x f))(x) \quad (5.21)$$

while the last equation should be understood in an algebraic way, as the function might not be differentiable and therefore the scale space description acts as a regularizer for the derivation.

Regarding the implementation details, it makes everything much easier, as we only have to compute the derivative of a Gaussian filter, which can be achieved by means of Difference of Gaussian [You87].

5.3.1.2 Canny edge detector

Another approach for scale dependent edge detection is known as the Canny edge detector [Can86] and its development as an Infinite Impulse Response filter: the Canny-Deriche variant [Der87].

To derive its edge detection operator, Canny described three criteria to be optimized by means of calculus of variations:

E1 an edge detection operator should be accurate; it means that the operator should detect all the edges and only edges,

E2 the detected edges should be well localised,

E3 each edge should be detected only once.

The last point is somehow contained in the previous ones; indeed, if an edge is being detected twice, one of them has to be considered as a false positive which yields a lower accuracy of the operator.

Based on these three criteria, the author derives functionals to be optimised. It comes out that the selection of an optimal operator depends on a tradeoff between localisation and performance (one getting better with the size of the operator and the second one getting better for smaller operator sizes).

5.3. MULTISCALE ANALYSIS AND DISTANCE TRANSFORMS

Defining an optimal operator can be done by solving an optimisation problem: maximising the product of the localisation and the performance, subject to a constraint about the distances between local maxima (this is for the criteria E3). Finally, this problem is solved by a sum of 4 exponential functions which can be approximated by Gaussian kernels. Edges are then found by a 4 step process:

1. a smoothing, getting rid of most of the noise,

2. a gradient computation, which yields both strength and orientation of edges,

3. a non maximum suppression, to avoid double detection of edges, and

4. an hysteresis thresholding, to enforce continuity of edges and local noisy detection.

The whole edge detection process is dependent on the size of the smoothing filter which determines the scale at which the features (edges or corners) are detected.

5.3.1.3 Comparison of scale dependent edge detections

We give here some examples of results achieved with both scale dependent edge detection algorithms. Moreover, we see how an appropriate choice of the different scale levels is important in designing a scaled distance transform.

Fig. 5.5 shows the two approaches on the Mansion image. It illustrates both edge detections at different scales and their distance transforms. Note that for visual reasons, images have been scaled from [minimum value-maximum value] to [0, 1].

As expected, the lowest scales give a high response to all textural components (as it can be seen on the left hand side of the figure, where the brick wall motif is appearing as edges) while the largest scales (to the right hand side of the figure) gives only the very coarse details of the image. We can mainly see the boundaries of the walls of the house and some part of the window, but no finer details. We should take care about such considerations when dealing with particular applications. For instance, if one wants to recognise whether two houses look globally the same, it might be interesting to give more impact to the lower frequency components (associated to the larger scales). On the opposite if one is interested in a textural comparison, then the higher frequency details do have their importance too.

When we look at the outputs of the distance transforms (second and last rows) at the lower scales, we can notice that it is hard to infer any global information out of their outputs. The small details covering very densely the image, the distance transform's range gets very low. On the other side, on the two most right columns, the global boundaries of the house are clearly visible.

5.3.2 Multiscale distance transform

5.3.2.1 Continuous distance transform on multiscale edges

We introduce here a novel distance transform representation of images. We base our approach by going back to the philosophy of the original work of Baddeley [Bad92] where the author actually wanted to compare edges.

In this new approach, we consider an image as a stack of binary edges at different scales, and compute a distance transform at each scale. This yields what we call the Scaled Distance Transform *SDT*.

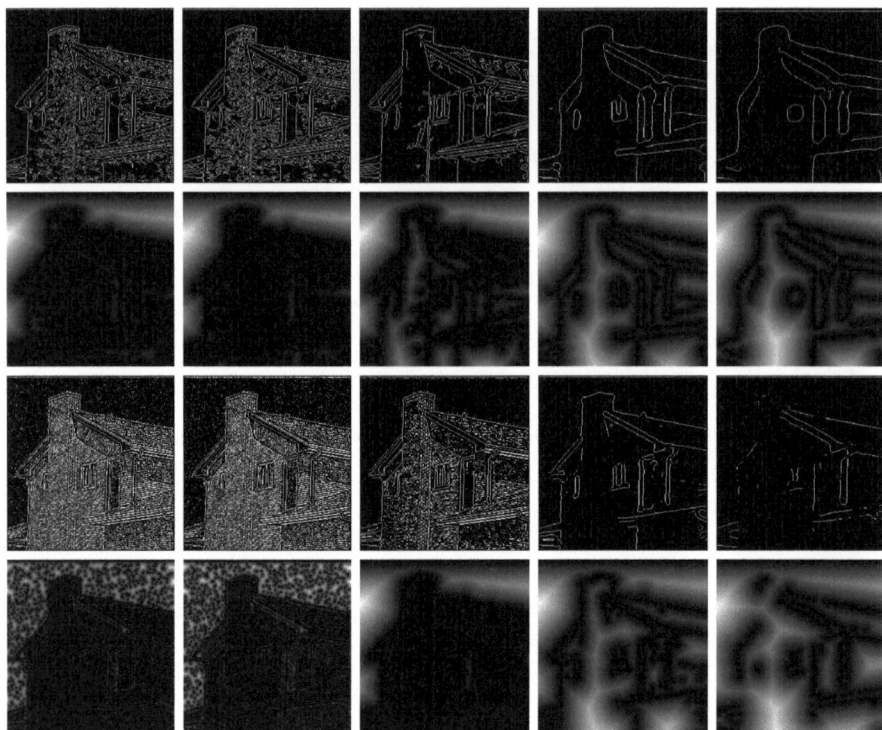

Figure 5.5: Effect of the different edge detections algorithms introduced given at different scales together with their distance transforms. The two first rows involve the Canny edge detector while the two last ones concerns the Gaussian scale space. First and third row are edges at different scales and second and last one are their distance transform (to the foreground). Scale increases from left to right.

Formally speaking, let us consider f an image defined on a bounded domain $D \subset \mathbb{R}^2$ with values in a set $G \subset \mathbb{R}$. We will in general have positive intensity values *i.e.* $G \subset \mathbb{R}^+$.

$$SDT(f)(x, y, s) = DT(E[f](s))(x, y), \tag{5.22}$$

with $E[f](s)$ corresponds to the edge detection of the image f at scale $s \geq 0$. As already said earlier, this can be done by computing the second derivative of Gaussian kernels, convolving the image with these kernels and finding zero-crossing. These edges are then used as foreground for the distance transform (when using the distance to the foreground variant, see Def. 15).

Note that for this representation to make sense, we need, at least, that the function is measurable. Then, the axioms of the scale spaces detailed in the previous section ensure that for all $s \geq 0$, $SDT(f)(\cdot, \cdot, s)$ is measurable.

Measuring differences In order to compare two images using our SDT, we introduce the following distance:

5.3. MULTISCALE ANALYSIS AND DISTANCE TRANSFORMS

Definition 26 (Metric between scaled distance transform). *Given two measurable functions f and g, we define*

$$d_{SDT}(f,g) = \int_{s \geq 0} \|SDT(f)(\cdot,\cdot,s) - SDT(g)(\cdot,\cdot,s)\| d\mu_s \qquad (5.23)$$

with μ_s being a finite and compact measure

The idea of having a compact measure, is that it suffices for the integral to converge. Interestingly, such representations can be seen as examples of classical Wiener spaces. If one assumes the scale to take values in an interval $[0,T]$ and for such an s we define $f_s := \phi[f,s]$ in a metric space $(\mathbb{R}^n, d)$, then we can apply the theory of classical Wiener spaces [Üst10]. This would allow a better understanding and theoretical motivations for such representations; however, this is left for future research as it would lead us far beyond the scope of this work.

In particular the measure μ_s will be responsible for giving importance to certain details of an image, or to certain sizes. More details will be given in the application part in a later section.

Using such image representation, we have also a monotonicity in the autocorrelation function with respect to translations

Property 7 (Monotonicity of the SDT's autocorrelation). *Consider $\Delta_{SDT}[f](t) := d_{SDT}(f, \tau_t f)$ the autocorrelation function based on the scaled distance transform representation, and consider comparing each scales with the Hausdorff distance. Then it holds*

$$\forall t, \forall 0 < \lambda_1 < \lambda_2, \Delta_{SDT}[f](\lambda_1 t) < \Delta_{SDT}[f](\lambda_2 t) \qquad (5.24)$$

Proof. Using the notation of observations introduced in the scale space analysis (see Eq. 5.18), we have

$$SDT(f)(x,s) = (DT(E[\phi[f,s]]))(x)$$

which means that

$$\Delta_{SDT}[f](t) = \int_{s \geq 0} H\left(SDT(f)(\cdot,\cdot,s) - SDT(\tau_t f)(\cdot,\cdot,s)\right) d\mu_s$$

and due to the translation invariance of both edge detection operation and scale space observation, it holds

$$\Delta_{SDT}[f](t) = \int_{s \geq 0} H\left(SDT(f)(\cdot,\cdot,s), \tau_t SDT(f)(\cdot,\cdot,s)\right) d\mu_s$$

so that for all $s \geq 0$, we compare two binary distance transforms which are shifted by t from one to another.

And now applying Prop. 5 concludes the proof.

□

5.3.2.2 Discrete distance transform on multiscale edges

We see here the applicability of our methods. The main difference with the continuous case is that the images are now defined on a pixel grid, and the intensities can only take a finite set of predefined values. Moreover, the space of scales will be discretised so that only a certain number of octaves are considered.

Formally speaking, f is considered as a discrete image defined on a domain $D \subset \mathbb{Z}^2$ with values in a set $G \subset \mathbb{Z}$. We will typically have $D = \{1, \cdots, N\} \times \{1, \cdots, M\}$ and $G = \{0, \cdots, 255\}$.

Measuring differences We review here some possibilities for measuring differences and introduce first a naive variant, which, as we will see in the application part, is not well suited for image comparison purposes. This norm gives the same role to both spatial coordinates and scale coordinate.

$$\Delta^{(global)}(SDT(f), SDT(g)) := \left(\frac{1}{|D| \cdot |S|} \sum_{x,y,s} |SDT(f)(x,y,s) - SDT(g)(x,y,s)|^p \right)^{1/p}. \quad (5.25)$$

However this representation lacks some finesse as it considers all dissimilarities at all levels exactly the same way. This may not be the most reasonable assumption one can do, as details at the finer level actually quite often correspond to noise or, in a more general case, to high frequency details (or local textures). If we go back to the metric introduce in the continuous case, we have the discrete equivalent:

$$\vec{\Delta}^{(scales)}[SDT(f), SDT(g)](s) := \left(\frac{1}{|D|} \sum_{x,y} |SDT(f)(x,y,s) - SDT(g)(x,y,s)|^p \right)^{1/p}. \quad (5.26)$$

These dissimilarities can then be combined in a weighted manner:

$$\Delta^{(scales)}(SDT(f), SDT(g)) := \frac{1}{N_\mu} \sum_{s \in S} \mu(s) \vec{\Delta}^{scales}[SDT(f), SDT(g)](s) \quad (5.27)$$

where N_μ stands for a normalization factor. It will generally be equal to $\sum_{s \in S} \mu(s)$.

As examples of such weighting functions μ we can either choose

- $\mu = 1_{s^*}$, for a certain $s^* \in S$, which is equivalent to considering a single scale for the edges; therefore not really interesting to us.

- $\mu = C$, where C denotes a positive constant. This case will give the same importance to each of the scales of the image. We will denote the dissimilarity measure based on this one as $\Delta^{(u)}$ (for uniform) in the followings.

- $\mu(s) = \alpha s, \forall s \in S$, where α denotes a positive real constant. This will give more emphasize on the details found on the larger scale (*i.e.* with s getting bigger). This dissimilarity will be denoted as $\Delta^{(p)}$ (for proportional) in the following.

5.3.2.3 Some examples

We want to now motivate the choice of such representations by analysing homogeneities in different databases. Most of the images can be found on the internet[1]. These images are particularly interesting

---

[1] Available from Patrick Gros' website on the IRISA:
http://www.irisa.fr/texmex/ressources/bases/base_images_movi/references.html

5.3. MULTISCALE ANALYSIS AND DISTANCE TRANSFORMS

in order to analyse how the structural similarities behave in presence of a certain kind of transformation. The first set of examples, Fig. 5.6, is taken as a database to assess the behaviour of the metrics against scale changes.

Figure 5.6: Example of images from three databases corrupted only by scaling. We refer to them as scaling sequences 1, 2 and 3.

The three sequences are depicted in inverted order of complexity. The first row shows a really cluttered scene with many details causing many changes at the pixel level. The second scene shows some big uniform areas with some local textures. The last scene shows mainly two regions, one containing nuts and the background. Both regions are rather uniform. Results are illustrated on the graphs in Fig. 5.7.

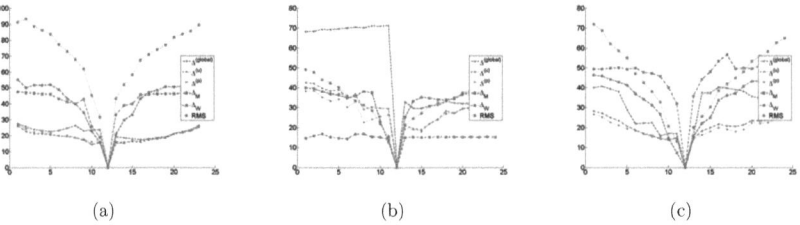

(a) (b) (c)

Figure 5.7: Behaviour of the different distance transform methods on the scaling sequences. All images are compared to a central one

As one can see, apart from Molchanov's approach distance transform methods are rather weak. Even the scale-based one, especially when used with the global measure. However, with the complexity of the image getting lower, the SDT framework manages to behaves almost monotonically with the scale changes. There is a lot to think that an optimised similarity measure could take advantage of a certain image.

The next set of images (some of them are illustrated in Fig. 5.8) is concerned with translation of a pattern. Some references can be seen together with some translations.

Figure 5.8: Example of images from three databases corrupted only by translation in the camera plane. They will be refered to as the translation sequences 1,2,3 and 4 respectively.

It should be noted that the two first sequences show some strong high frequency components while the two last one are better described by big structures. We should expect to see this characteristic on the experiments.

Fig. 5.9 shows some results of the different distance transform-based metrics when facing translation in the camera plane. Ideally it should behaves linearly with the amount of shift. It is hard to tell this at the moment, as we can only assess visually that the translation is almost constant from an image to the other. However, under this assumption, we can notice that most of the metrics have a rather monotonic behaviour, at least near the optimal point. As we were expecting it, the distance transforms are rather weak on the more textured sequences (sequences 1 and 2) compared to the results for more smooth images (sequences 3 and 4).

When analysing the results of the two first sequences, we can see that all distance transform-based metrics tend to become stable to a certain value very soon. In our opinion, this is due to the highly

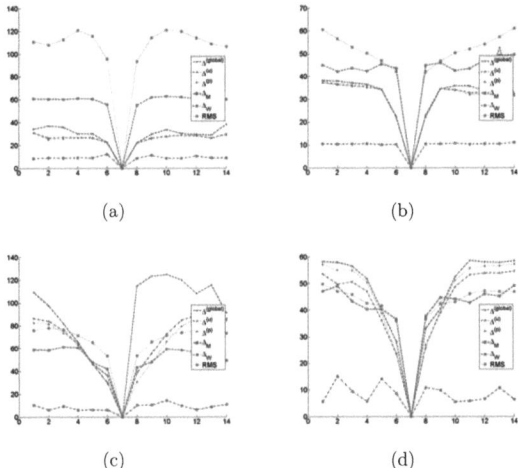

Figure 5.9: Behaviour of the different distance transform methods on the translation sequences. All images are compared to a central one

cluttered images which give many edges at the high frequency components. Therefore, no valuable information is kept from an image to the other.

On both sequences of smooth images, the SDT approach combined with the proportional μ ($\Delta^{(p)}$) behaves very well while the $\Delta^{(global)}$ metric does not give valuable information about the comparisons.

Note that the RMS coefficient as been scaled by 512 (size of the image) and that both Wilson and Molchanov metrics have been multiplied by 10 (chosen so that the scales fit more or less with the other ones')

Finally, a last example for such simple transformations deals with illumination changes. In Fig. 5.10 an image is being saved with different illumination intensities, gradually from darker to lighter images.

These three scenes show a good mix between high frequency and lower frequency information. Once again the different measures have been computed and results are depicted on Fig. 5.11.

As a natural result, the RMS score is monotonic and behaves nicely. This is due to the contrast changes being modeled by a simple addition/subtraction of intensities. Molchanov's coefficient as well as the $\Delta^{(u)}$ and $\Delta^{(p)}$ scores show a monotonic behaviour even though they are not linear. On the opposite it seems that Wilson's score and our $\Delta^{(global)}$ metric are not well suited to handle illumination changes.

Analysis As a conclusion of the previous results, we can remark that the method introduced above shows some interesting properties regarding simple tranformations. Yet a better definition of the comparison score needs to be done to handle local characteristics of the image. For instance, when dealing with intensity changes, it would be interesting to adapt the weighting factor to the local intensity level. Moreover as it has been proved earlier, and as we can see on the results depicted on Fig. 5.9, the SDT approach allows to give some good insights about the shift between two images.

Another point to be noted is the poor results of Wilson's coefficient when considering highly cluttered scenes. This might be due to a wrong parameterisation of the algorithm (for instance,

Figure 5.10: Example of images from three databases corrupted only by illumination changes. We refer to them as illumination sequences 1, 2, and 3.

dealing with other weighting factors, or with a finer discretisation of the illumination scale).

Finally scale changes seem to be the hardest to get and only Molchanov's metric manages to get the essence of the distortions. However, an adaptive scale space should be able to handle this and could be included in the process of the SDT.

5.3. MULTISCALE ANALYSIS AND DISTANCE TRANSFORMS

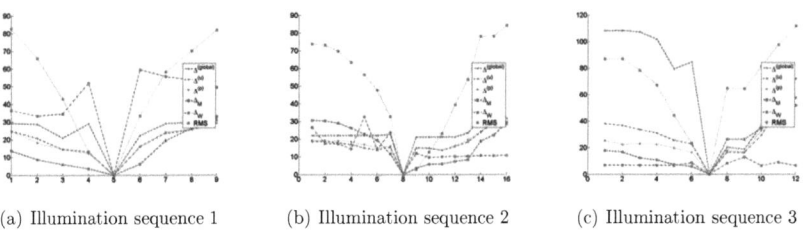

(a) Illumination sequence 1   (b) Illumination sequence 2   (c) Illumination sequence 3

Figure 5.11: Behaviour of the different distance transform methods on the illumination sequences. All images are compared to a central one

Chapter 6

An analytic approach to structures

6.1 Elements of the theory of analytic signals

In his pioneer work, Gabor [Gab46] was interested in efficient and reliable communication or transmission of information. He therefore introduced a novel representation of signals based on a mix of time and frequency analysis motivated by the fact that information cannot be well localized in both its frequency and time. To this end he needed to introduce complex representations of signals based on odd and even parts of *quadrature filters* obtained by means of the Hilbert transform. It led him to introduce time frequency analysis on what he called *logon*, an "elementary quantum of information" It can be shown that such representations are optimal in the sense that they can convey the maximum amount of information for a certain duration or that they need the smallest time duration for a certain amount of information.

Later, Ville [Vil48] was interested in the energy transportation in a network. He used such Hilbert transformed representations to introduce local characteristics of a signal, as we will recall in Sec. 6.1.4

While the initial works were mainly involved in transmission and coding of one dimensional information, such Gaussian weighted harmonics have been analyzed and extended to higher dimensional signals to give raise to what is now known as Gabor filters [Dau85, Dau88b] in image processing for instance, and other complex analytic representation [Hah92]

This particular role of the Hilbert transform in the design for quadrature filters (which will not be detailed in this book as it goes far beyond its scope), makes it a very studied tool in signal and image processing (see for instance [SA90, FA91, AKM95, MSRV97, MFS97, BNB04] for a non-exhaustive results).

6.1.1 The Hilbert transform

6.1.1.1 Fourier transforms

Before we go further, we recall the definitions of the Fourier transform that we are going to use within this chapter. Moreover, some properties are given to the reader as general introduction

Definition 27 (Fourier transforms). *Let $f \in L^1(\mathbb{R}, \mathbb{R})$, the Fourier transform can be defined as*

$$\mathcal{F}(f)(\xi) := \widehat{f}(\xi) = \int_{-\infty}^{\infty} f(x) e^{-i 2\pi x \xi} \mathrm{d}x \qquad (6.1)$$

Using a density argument this definition can be extended to any $f \in L^2$.
The inverse Fourier transform is given as

$$\mathcal{F}^{-1}(f)(x) := \int_{-\infty}^{\infty} \widehat{f}(\xi) e^{i 2\pi x \xi} \mathrm{d}\xi \qquad (6.2)$$

It should be noticed that in general $\mathcal{F}^{-1}(\mathcal{F}(f)) \neq f$. However, both members of the inequality are equal when the function f is continuous.

Remarking that for L^2 functions, with $\langle \cdot, \cdot \rangle$ being the standard scalar product, it holds (Plancherel theorem)

$$\langle \widehat{f}, g \rangle = \langle f, \widehat{g} \rangle \qquad (6.3)$$

We can therefore extend this definition to the space $\mathcal{S}'$ of tempered distributions [Sch61] ($\mathcal{S}'$ is the dual space of $\mathcal{S} = \{f \in C^\infty(\mathbb{R}^n, \mathbb{C}) : \forall \alpha, \beta, \text{ multiindices}, \sup_{x \in \mathbb{R}^n} |x^\alpha D^\beta(x)| < \infty\}$)[1]: for $f \in \mathcal{S}$, its Fourier transform is defined as the tempered distribution $\widehat{f}$ such that

$$\forall \phi \in C^\infty \langle \widehat{f}, \phi \rangle = \langle f, \widehat{\phi} \rangle \qquad (6.4)$$

As an example, let us consider the case of the Dirac δ distribution:

$$\langle \widehat{\delta}, \phi \rangle = \langle \delta, \widehat{\phi} \rangle = \widehat{\phi}(0)$$
$$= \int_{-\infty}^{\infty} \phi(x) e^{i 2\pi x \cdot 0} \mathrm{d}x = \langle 1, \phi \rangle \qquad (6.5)$$

and therefore, we can write $\widehat{\delta} = 1$.

This example is very important when used with the rules of Fourier calculus. For instance, we have, by applying two times the Fourier transform, the Delta distribution being real, that $\widehat{1} = \delta$. If now apply the Fourier shift theorem, we get that $\widehat{e^{i\xi_0 x}}(\xi) = \delta(\xi - \xi_0)$. We need this result in a later section when computing some examples of analytic signals.

6.1.1.2 Hilbert transforms

Definition 28 (Hilbert Transform). *The Hilbert transform of a signal $f \in L^2(\mathbb{R})$ (or more generally $f \in L^p(\mathbb{R}), 1 < p < \infty$) is defined either in the spatial domain as a convolution with the Hilbert kernel or as a Fourier multiplier:*

$$\mathcal{H} f = h * f \qquad (6.6)$$
$$\mathcal{F}(\mathcal{H}f)(\xi) = -i \operatorname{sign}(\xi) \mathcal{F}(f)(\xi) \qquad (6.7)$$

where we have used two functions:

- fhe Hilbert kernel $h(x) = \frac{1}{\pi x}$,
- the operator $\operatorname{sign}(\xi) = \begin{cases} 1 & \xi > 0 \\ 0 & \xi = 0 \\ -1 & \xi < 0 \end{cases}$

---
[1] *i.e.* the set of all linear functionals on $\mathcal{S}$

6.1. ELEMENTS OF THE THEORY OF ANALYTIC SIGNALS

Using the Cauchy principal value of a function with singularity at 0 (there are some other equivalent definitions, in particular, when the singularity is located elsewhere):

$$\mathcal{P.V.} f = \lim_{\varepsilon \to 0^+} \left\{ \int_{-\infty}^{-\varepsilon} f(x)\mathrm{d}x + \int_{\varepsilon}^{\infty} f(x)\mathrm{d}x \right\} = \lim_{\varepsilon \to 0^+} \int_{|x| > \varepsilon} f(x)\mathrm{d}x \qquad (6.8)$$

we can now prove that the expression of the Hilbert transform as a Fourier multiplier 6.7 indeed holds. We define the Hilbert transform as the principal value of the Hilbert kernel in the sense of distributions:

$$\langle \mathcal{P.V.} \frac{1}{x}, \phi \rangle = \lim_{\varepsilon \to 0} \int_{|x|>\varepsilon} \frac{\phi(x)}{x}\mathrm{d}x \qquad (6.9)$$

$$\langle \widehat{\mathcal{P.V.} \frac{1}{x}}, \phi \rangle = \langle \mathcal{P.V.} \frac{1}{x}, \widehat{\phi} \rangle = \lim_{R \to \infty} \int_{-R}^{R} \frac{1}{2} \frac{\widehat{\phi}(x) - \widehat{\phi}(-x)}{x}\mathrm{d}x \qquad (6.10)$$

$$= \lim_{R \to \infty} \int_{-R}^{R} \frac{1}{2} \frac{\int_{\mathbb{R}} \phi(\xi) e^{-i2\pi x\xi} \mathrm{d}\xi - \int_{\mathbb{R}} \phi(\xi) e^{i2\pi x\xi} \mathrm{d}\xi}{x}\mathrm{d}x \qquad (6.11)$$

$$= \lim_{R \to \infty} \int_{-R}^{R} \int_{\mathbb{R}} \frac{1}{2} \frac{\phi(\xi)\left(e^{-i2\pi x\xi} - e^{i2\pi x\xi}\right)}{x}\mathrm{d}\xi \mathrm{d}x \qquad (6.12)$$

$$= -i \lim_{R \to \infty} \int_{\mathbb{R}} \phi(\xi) \int_{-R}^{R} \frac{\sin 2\pi x\xi}{x}\mathrm{d}x \mathrm{d}\xi = -i \int_{\mathbb{R}} \phi(\xi) \lim_{R \to \infty} \int_{-R}^{R} \frac{\sin 2\pi x\xi}{x}\mathrm{d}x \mathrm{d}\xi \qquad (6.13)$$

The last row can be computed using two times the result of the Dirichlet integral (result obtained using the theorem of the residuals in complex analysis [Rud70])

$$\int_0^\infty \frac{\sin(t)}{t}\mathrm{d}t = \frac{\pi}{2} \qquad (6.14)$$

applied to the previous line gives

$$\langle \widehat{\mathcal{P.V.} \frac{1}{x}}, \phi \rangle = -i \int_{\mathbb{R}} \phi(\xi) \left(\int_{-\infty}^{0} \frac{\sin 2\pi x\xi}{x}\mathrm{d}x + \int_{0}^{\infty} \frac{\sin 2\pi x\xi}{x}\mathrm{d}x \right) \mathrm{d}\xi \qquad (6.15)$$

$$= -i \int_{\mathbb{R}} \phi(\xi) \left(\int_{-\mathrm{sign}(\xi)\infty}^{0} \frac{\sin t}{t}\mathrm{d}t + \int_{0}^{\mathrm{sign}(\xi)\infty} \frac{\sin t}{t}\mathrm{d}t \right) \mathrm{d}\xi \qquad (6.16)$$

$$= -i \int_{\xi \in \mathbb{R}^+} \phi(\xi) \left(\int_{-\infty}^{0} \frac{\sin t}{t}\mathrm{d}t + \int_{0}^{\infty} \frac{\sin t}{t}\mathrm{d}t \right) \mathrm{d}\xi \qquad (6.17)$$

$$+ -i \int_{\xi \in \mathbb{R}^-} \phi(\xi) \left(\int_{\infty}^{0} \frac{\sin t}{t}\mathrm{d}t + \int_{0}^{-\infty} \frac{\sin t}{t}\mathrm{d}t \right) \mathrm{d}\xi \qquad (6.18)$$

$$= -i \int_{\xi \in \mathbb{R}^+} \phi(\xi)\pi \mathrm{d}\xi + i \int_{\xi \in \mathbb{R}^+} \phi(\xi)\pi \mathrm{d}\xi \qquad (6.19)$$

$$= -i \int_{\xi \in \mathbb{R}} \pi \, \mathrm{sign}(\xi) \phi(\xi) \mathrm{d}\xi \qquad (6.20)$$

$$= \langle -i\pi \, \mathrm{sign}, \phi \rangle \qquad (6.21)$$

which means that the Hilbert transform can be written as a Fourier multiplier with symbol

$$\widehat{\mathcal{P}.\mathcal{V}.\frac{1}{\pi x}} = -i\,\text{sign}.$$

6.1.2 The analytic signal representation

With the previous theoretical aspect in mind, we can now introduce the concept of analytic signal.

Definition 29 (Analytic Signal)**.** *The analytic signal is computed as a complex combination of both the original signal and its Hilbert transform:*

$$f_A = f + i\mathcal{H}f \qquad (6.22)$$

6.1.3 Properties of the Hilbert transform and analytic signal representation

Property 8 (Properties of the Hilbert Transform)**.** *Given a signal f the followings hold true:*

- $\forall \xi \neq 0,\ |\widehat{\mathcal{H}f}(\xi)| = |\mathcal{F}(f)(\xi)|$

- $\mathcal{H}\mathcal{H}f = -f \Rightarrow \mathcal{H}^{-1} = -\mathcal{H}$

Proof. First remark is clear. $(\forall \xi,\ |\mathcal{H}f(\xi)| = |-i\,\text{sign}(\xi)\mathcal{F}(f)(\xi)| = |\mathcal{F}(f)(\xi)|).$

Second assertion makes use of the Hilbert transform on the Hilbert transformed signal after noticing that

$$\mathcal{H} = \mathcal{F}^{-1} \circ (-i\,\text{sign}(\cdot)) \circ \mathcal{F}$$

which gives

$$\mathcal{H}(\mathcal{H}(f)) = \mathcal{F}^{-1} \circ \{(-i\,\text{sign}(\cdot))\}^2 \circ \mathcal{F}(f)$$

Noticing that $\{(-i\,\text{sign}(\cdot))\}^2 = -1$ finishes the proof. □

Following its definition, we can notice that the Hilbert transform acts as a phase shifting of the original signal: if we write $i = e^{i\pi/2}$, the phase of the Fourier spectrum of the Hilbert transformed signal is obtained after a rotation of $\pm 90°$ of the phase of the original signal.

We can also remark, that due to the definition of the sign function, the Hilbert transform deletes the DC component and we should be careful in practical application to always have zero-mean signals (it is anyway formally not defined on such signals).

We have the following properties

Property 9.

$$\langle f, \mathcal{H}f \rangle_{L^2} = 0 \ \ Orthogonality \qquad (6.23)$$
$$\|f\|_2^2 = \|\mathcal{H}f\|_2^2 \ \ Energy \qquad (6.24)$$

6.1. ELEMENTS OF THE THEORY OF ANALYTIC SIGNALS

Proof. Orthogonality

$$\begin{aligned}
\langle \mathcal{H}f, f \rangle &= \int_{-\infty}^{\infty} \overline{\mathcal{H}f(x)} f(x) dx = \int_{-\infty}^{\infty} \overline{-i\frac{\xi}{|\xi|}\widehat{f}(\xi)} \widehat{f}(\xi) d\xi \\
&= \int_{-\infty}^{\infty} i\frac{\xi}{|\xi|} \overline{\widehat{f}(\xi)} \widehat{f}(\xi) d\xi = \int_{-\infty}^{\infty} i\frac{\xi}{|\xi|} |\widehat{f}(\xi)|^2 d\xi \\
&= i \left\{ \int_{-\infty}^{0} -|\widehat{f}(\xi)|^2 d\xi + \int_{0}^{\infty} |\widehat{f}(\xi)|^2 d\xi \right\} \\
&= i \left\{ \int_{\infty}^{0} |\widehat{f}(-t)|^2 dt + \int_{0}^{\infty} |\widehat{f}(\xi)|^2 d\xi \right\} = 0
\end{aligned}$$

due to the fact that we are dealing with real valued signals.

Energy

$$\begin{aligned}
\|\mathcal{H}f\|_2^2 &= \int_{-\infty}^{\infty} |\mathcal{H}f(x)|^2 dx \\
(\text{Plancherel}) &= \int_{-\infty}^{\infty} |\widehat{\mathcal{H}f}(\xi)|^2 d\xi \\
&= \int_{-\infty}^{\infty} |-i\frac{\xi}{|\xi|}\widehat{f}(\xi)|^2 d\xi \\
&= \int_{-\infty}^{\infty} |\widehat{f}(\xi)|^2 d\xi = \|f\|_2^2
\end{aligned}$$

□

This description of the analytic signal gives interesting property regarding its Fourier spectrum:

Property 10. *The Fourier spectrum of an analytic signal is one sided and has an amplitude (on the positive side) twice as big as the original signal's spectrum's one.*

Proof. Assume f_A is an analytic signal, $\exists f$, such that, $f_A = f + i\mathcal{H}f$. By computing its Fourier transformation (which is linear) we get $\mathcal{F}(f_A)(\xi) = \mathcal{F}(f)(\xi) + i \times (-i\,\text{sign}(\xi)\mathcal{F}(\mathcal{H}f)(\xi))$. Now if we apply Eq. 6.7, we get $\mathcal{F}(f_A)(\xi) = (1 + \text{sign}(\xi))\mathcal{F}(f)(\xi)$.

Therefore, on the negative part of the spectrum $(\text{sign}(\xi) < 0)$, we have $1 + \text{sign}(\xi) = 0$ while on the positive part, we get $1 + \text{sign} = 2$ and it has only a factor 1 for $\xi = 0$ according to the definition of the sign function, which proves the property. □

6.1.4 Analytic signal analysis

Local features computation

Definition 30 (Local features).

$$A(x) = \sqrt{f(x)^2 + \mathcal{H}f(x)^2} \tag{6.25}$$

$$\varphi(x) = \arctan\left(\frac{\mathcal{H}f(x)}{f(x)}\right) = \arctan\left(\frac{\Im(f_A(x))}{\Re(f_A(x))}\right) \tag{6.26}$$

In this case, the arctan function should have values in $[0, 2\pi)$ or $[-\pi, \pi)$ or any interval of length 2π when considering the signs of the different components in the four quadrants of the complex plane.

Property 11 (Invariance - equivariance, Split of identity [FS01]). *The local phase together with the local amplitude fulfil the property of invariance-equivariance:*

- *The local phase depends only on the local structure*

- *The local amplitude depends only on the local energy*

If moreover these features are a complete description of the signal, they are said to perform a split of identity.

An application to AM-FM signal demodulation We want to show here on an easy example how such an analytic representation can be interesting for signal analysis. Assume we have a signal f of the spatial or time variable x which is modulated in its amplitude A and its phase φ. We can for instance write

$$f(x) = A(x)\cos(\varphi(x))$$

We want to recover independently both amplitude and phase functions. We can moreover assume A is a slowly changing function while the phase φ is highly oscillating. We can for instance consider $A(x) = \frac{1}{2}\cos(\omega_A x + \varphi_{A,0}) + 1$ and $\varphi(x) = \omega_\varphi x + \varphi_0$ with $\omega_A \ll \omega_\varphi$.

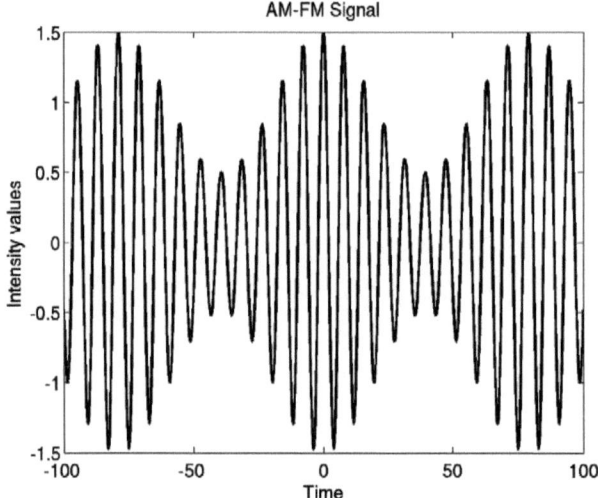

Figure 6.1: An example of amplitude and frequency modulated signal

We can now apply the local features analysis introduced above which yields

$$A(x) = |f(x) + i\mathcal{H}f(x)| = \frac{1}{2}\cos(\omega_A x + \phi_{A,0}) + 1$$
$$\varphi(x) = Atan\left(\frac{\mathcal{H}f(x)}{f(x)}\right) = \omega_\phi x + \phi_0 \text{ mod } 2\pi$$

6.1. ELEMENTS OF THE THEORY OF ANALYTIC SIGNALS

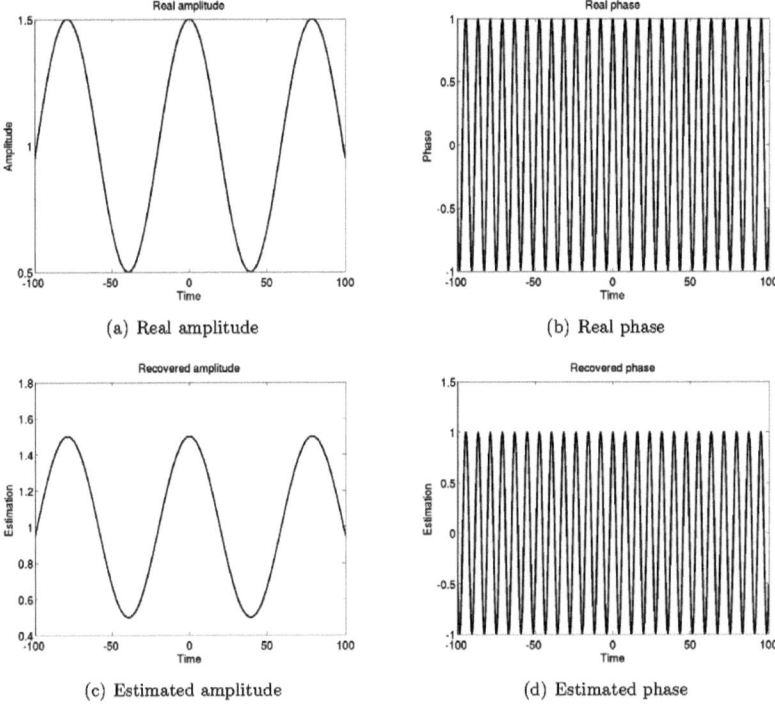

Figure 6.2: Original and recovered amplitude and phase of an AM-FM signal.

6.1.5 The analytic signal as a boundary value problem in complex analysis

While the analytic signal is a very common concept in the field of signal theory, its basic mathematics can be derived from the theory of analytic functions [FS01, BBRH13]. The close connection can be understood when considering the following Riemann-Hilbert problem with respect to the complex parameter $z = x + iy$:

$$\frac{\partial F}{\partial \bar{z}}(z) = 0 \quad z \in \mathbb{C}, y \geq 0, \tag{6.27}$$

$$\Re(F(x)) = f(x) \quad x \in \mathbb{R}. \tag{6.28}$$

One solution of this problem is given by the Cauchy integral

$$F(z) = F_\Gamma f(z) := \frac{1}{2\pi i} \int_\mathbb{R} \frac{1}{\tau - z} f(\tau) d\tau. \tag{6.29}$$

Of course this solution is unique only up to a constant. Normally, this constant will be fixed by the condition $\Im(F(z_0)) = c$, i.e. the imaginary part of F given in an interior point.

When we now consider the trace of F_Γ, i.e. the boundary value for $\Im(z) = 0$, we arrive at the so-called Plemelj-Sokhotzki formula:

$$\mathrm{tr} F_\Gamma f = \frac{1}{2}(I + i\mathcal{H})f = \frac{1}{2}f + \frac{1}{2}i\mathcal{H}f =: P_\Gamma f. \tag{6.30}$$

Up to the factor 1/2 this corresponds to our above definition of an analytic signal.

In this way an analytic signal represents the boundary values of an analytic function in the upper half plane (or for periodic functions in the unit disc). Starting from this concept we are going now to take a look at higher dimensional generalizations.

6.2 Generalizations to higher dimensions

The generalization of the concept of analytic signal to higher dimensional spaces is not straightforward. We describe [BBRH13] here two approaches based respectively on multivariable complex analysis or Clifford analysis.

6.2.1 The higher dimensional analytic signal

6.2.1.1 A single orthant description

In 1998 Bülow proposed a definition of a hypercomplex signal based on the so-called partial and total Hilbert transforms [BS01], these concepts are more properly defined later in this section. To see that analytic signals are boundary values of functions in a Hardy space[2] we consider the following Riemann-Hilbert problem for two complex variables $z_1 = x_1 + iy_1$ and $z_2 = x_2 + iy_2$ in $\mathbb{C}^2$:

$$\frac{\partial F}{\partial \bar{z}_1}(z_1, z_2) = 0 \qquad (z_1, z_2) \in \mathbb{C}^2, y_1, y_2 \geq 0, \tag{6.31}$$

$$\frac{\partial F}{\partial \bar{z}_2}(z_1, z_2) = 0 \qquad (z_1, z_2) \in \mathbb{C}^2, y_1, y_2 \geq 0, \tag{6.32}$$

$$\Re(F(x_1, x_2)) = f(x_1, x_2) \qquad x_1, x_2 \in \mathbb{R}^2. \tag{6.33}$$

For the solution, (see e.g. [Dzh96] or [Rud80]), we just want to point out that the domain is a polydomain in the sense of $\mathbb{C}^n$, so that we can give it in form of the Cauchy integral:

$$F(z_1, z_2) = \frac{1}{4\pi^2} \int_{\mathbb{R}^2} \frac{1}{(\xi_1 - z_1)(\xi_2 - z_2)} f(\xi_1, \xi_2) \mathrm{d}\xi_1 \mathrm{d}\xi_2. \tag{6.34}$$

Now again looking at the corresponding Plemelj-Sokhotzki formula we get

$$\begin{aligned}\mathrm{tr} F(x_1, x_2) = &\ \tfrac{1}{4}f(x_1, x_2) - \tfrac{1}{4}\int_{\mathbb{R}^2} \tfrac{1}{(\xi_1-x_1)(\xi_2-x_2)} f(\xi_1, \xi_2) \mathrm{d}\xi_1 \mathrm{d}\xi_2 \\ &+ i\tfrac{1}{4}\left(\int_{\mathbb{R}} \tfrac{1}{\xi_1-x_1} f(\xi_1, x_2) \mathrm{d}\xi_1 + \int_{\mathbb{R}} \tfrac{1}{\xi_2-x_2} f(x_1, \xi_2) \mathrm{d}\xi_2\right)\end{aligned} \tag{6.35}$$

which up to the factor 1/4 corresponds to the definition of an analytic signal by Hahn [Hah92]. Here

---

[2]Hardy spaces are sets of analytic functions in complex analysis or their boundary values in the context of real analysis

6.2. GENERALIZATIONS TO HIGHER DIMENSIONS

$$\mathcal{H}_1 f = \int_{\mathbb{R}} \frac{1}{\xi_1 - x_1} f(\xi_1, \cdot) \mathrm{d}\xi_1 \tag{6.36}$$

$$\mathcal{H}_2 f = \int_{\mathbb{R}} \frac{1}{\xi_2 - x_2} f(\cdot, \xi_2) \mathrm{d}\xi_2 \tag{6.37}$$

are called the partial Hilbert transforms and

$$\mathcal{H}_T f = \frac{1}{4} \int_{\mathbb{R}^2} \frac{1}{(\xi_1 - x_1)(\xi_2 - x_2)} f(\xi_1, \xi_2) \mathrm{d}\xi_1 \mathrm{d}\xi_2 \tag{6.38}$$

the total Hilbert transform. On the level of Fourier symbols we get

$$\mathcal{F}(\mathrm{tr}F)(\xi_1, \xi_2) = (1 + \mathrm{sign}\xi_1)(1 + \mathrm{sign}\xi_2) \mathcal{F} f(\xi_1, \xi_2). \tag{6.39}$$

Let us now take a look at the definition of Bülow. To this end we consider F to be a function of two variables z_1 and $\mathfrak{z}_2$ with two different imaginary units i and j (with $i^2 = j^2 = -1$), i.e. $z_1 = x_1 + iy_1$ and $\mathfrak{z}_2 = x_2 + jy_2$. We remark that both imaginary units can be understood as elements of the quaternionic basis with multiplication rules $ij = -ji = k$. In this way the above Riemann-Hilbert problem can be rewritten as

$$\frac{\partial}{\partial \overline{z_1}} F(z_1, \mathfrak{z}_2) = 0 \quad (z_1, \mathfrak{z}_2) \in \mathbb{C}^2, y_1, y_2 \geq 0, \tag{6.40}$$

$$F \frac{\partial}{\partial \overline{\mathfrak{z}_2}}(z_1, \mathfrak{z}_2) = 0 \quad (z_1, \mathfrak{z}_2) \in \mathbb{C}^2, y_1, y_2 \geq 0, \tag{6.41}$$

$$\Re(F(x_1, x_2)) = f(x_1, x_2) \quad x_1, x_2 \in \mathbb{R}^2, \tag{6.42}$$

where the second equation should be understood as $\partial_{\overline{\mathfrak{z}_2}}$ being applied from the from the right due to the non-commutativity of the complex units i and j.

The solutions is given by

$$F(z_1, \mathfrak{z}_2) = \frac{1}{4\pi^2} \int_{\mathbb{R}^2} \frac{1}{(\xi_1 - z_1)(\xi_2 - \mathfrak{z}_2)} f(\xi_1, \xi_2) \mathrm{d}\xi_1 \mathrm{d}\xi_2 \tag{6.43}$$

so that we get from the Plemelj-Sokhotzki formulae

$$\mathrm{tr}F(x_1, x_2) = \frac{1}{4}(I + i\mathcal{H}_1)(I + j\mathcal{H}_2) f(x_1, x_2) \tag{6.44}$$

$$= \frac{1}{4}(f + i\mathcal{H}_1 f + j\mathcal{H}_2 f + k\mathcal{H}_T f)(x_1, x_2). \tag{6.45}$$

While this is now a quaternionic-valued function, it still corresponds to a boundary value of a function holomorphic in two variables. For the representation in Fourier domain, one has to keep in mind that now the Fourier transform is to be applied another way: with respect to the complex plane in i and again with respect to the complex plane generated by j. Taking into account that $ij = -ji$ one arrives at the so-called quaternionic Fourier transform [Hit07, BS01]:

$$\mathcal{QF} f = \int_{\mathbb{R}^2} e^{ix_1 \xi_1} f(x_1, x_2) e^{jx_2 \xi_2} \mathrm{d}x_1 \mathrm{d}x_2, \tag{6.46}$$

and the following representation as Fourier symbols

$$\mathcal{QF}(\mathrm{tr}F)(\xi_1,\xi_2) = (1+\mathrm{sign}\xi_1)(1+\mathrm{sign}\xi_2)\mathcal{QF}f(\xi_1,\xi_2) \qquad (6.47)$$

In his early work, Hahn [Hah92] mentioned this approach as a single-orthant representation. To understand this, let us have a look at the Fourier spectrum of the multidimensional analytic signal. As we have seen earlier, its Fourier symbol reads $(1+\mathrm{sign}\xi_1)(1+\mathrm{sign}\xi_2)\mathcal{F}f(\xi_1,\xi_2)$ which we write $\mathcal{F}f(\xi_1,\xi_2)(1+\mathrm{sign}\,\xi_1+\mathrm{sign}\,\xi_2+\mathrm{sign}\,\xi_1\mathrm{sign}\,\xi_2)$ so that we have 4 different cases:

$$\begin{aligned}\xi_1>0 & \quad \xi_2>0 & \mathcal{F}(f_A)(\xi_1,\xi_2)=4\mathcal{F}(f)(\xi_1,\xi_2),\\ \xi_1>0 & \quad \xi_2<0 & \mathcal{F}(f_A)(\xi_1,\xi_2)=0,\\ \xi_1<0 & \quad \xi_2>0 & \mathcal{F}(f_A)(\xi_1,\xi_2)=0,\\ \xi_1<0 & \quad \xi_2<0 & \mathcal{F}(f_A)(\xi_1,\xi_2)=0,\end{aligned}$$

and therefore, we can notice that the Fourier transform of an analytic representation in higher dimension is fully contained in the domain with positive coordinates whence the so-called single-orthant definition.

6.2.1.2 Multidimensional analytic signal analysis

In image analysis problems, according to [Hah92] we can introduce the following features:

Local amplitude The local amplitude of a multidimensional analytic signal is defined in a similar way as for the one-dimensional case:

$$A_A(x,y) = \sqrt{|f(x,y)|^2 + |\mathcal{H}_1 f(x,y)|^2 + |\mathcal{H}_2 f(x,y)|^2 + |\mathcal{H}_T f(x,y)|^2}. \qquad (6.48)$$

This is also denoted as *energetic information*.

Local phase The phase is a feature describing how much a vector or quaternion number diverge from the real axis. It is defined in a similar manner as for the classical complex plane.

$$\varphi_A = \arctan\left(\frac{\sqrt{\mathcal{H}_1 f^2 + \mathcal{H}_2 f^2 + \mathcal{H}_T f^2}}{f}\right) \qquad (6.49)$$

This angle φ_A is what is denoted as phase or *structural information*.

Local orientation As we are at the moment interested in 2D signals (=images), we can also describe an orientation information, as the principal direction carrying the phase information. The imaginary plane, spanned by $\{i,j\}$, is two-dimensional and therefore we can also define an angle θ_A in this plane:

$$\theta_A = \arctan\left(\frac{\mathcal{H}_2 f}{\mathcal{H}_1 f}\right) \qquad (6.50)$$

This new angle is called the orientation of the signal or *geometric information*.

6.2.2 The monogenic signal

Another approach towards generalizing the analytic signal to higher dimensions is based on Clifford analysis.

6.2.2.1 A boundary value problem in Clifford analysis

We derive here the similar boundary value problems based on Clifford analysis instead of multidimensional complex analysis [BBRH13]. We see that similar results can be obtained by solving a Riemann-Hilbert problem on Dirac operator and how this leads to the monogenic representation of felsberg and Sommer [FS01].

Here we use a so-called Clifford algebra $C\ell_{0,n}$ [BDS82]. This is the free algebra constructed over $\mathbb{R}^n$ generated modulo the relation

$$x^2 = -|x|^2 \mathbf{e}_0 \quad x \in \mathbb{R}^n \tag{6.51}$$

where $\mathbf{e}_0$ is the identity of $C\ell_{0,n}$. For the algebra $C\ell_{0,n}$ we have the anti-commutation relationship

$$\mathbf{e}_i \mathbf{e}_j + \mathbf{e}_j \mathbf{e}_i = -2\delta_{ij} \mathbf{e}_0, \tag{6.52}$$

where δ_{ij} is the Kronecker symbol. Each element x of $\mathbb{R}^n$ may be represented by

$$x = \sum_{i=1}^n x_i \mathbf{e}_i. \tag{6.53}$$

A first-order differential operator which factorizes the Laplacian is given as the so-called Dirac operator

$$Df(x) = \sum_{j=1}^n \frac{\partial f}{\partial x_j}. \tag{6.54}$$

The Riemann-Hilbert problem for the Dirac operator can be stated in the form

$$DF(x) = 0 \quad x \in \mathbb{R}^3, x_3 > 0 \tag{6.55}$$

$$\Re(F(x_1, x_2)) = f(x_1, x_2, 0) \quad x_1, x_2 \in \mathbb{R}^2 \tag{6.56}$$

To solve this problem we follow the same idea as in the multivariable complex case.

$$F_\Gamma f = \int_{\mathbb{R}^2} \frac{x-y}{|x-y|^2} \mathbf{e}_3 f(x_1, x_2) \mathrm{d}x_1 \mathrm{d}x_2 \tag{6.57}$$

$$\mathrm{tr} F_\Gamma f = \frac{1}{2}(I + S_\Gamma)f = \frac{1}{2} f(\tilde{y}_1, \tilde{y}_2)$$
$$+ \frac{1}{2} \int_{\mathbb{R}^2} \frac{\mathbf{e}_1(x_1 - \tilde{y}_1) + \mathbf{e}_2(x_2 - \tilde{y}_2)}{|x-y|^2} \mathbf{e}_3 f(x_1, x_2) \mathrm{d}x_1 \mathrm{d}x_2. \tag{6.58}$$

Because the quaternions $\mathbb{H}$ are isomorphic to the even subalgebra $C\ell_{0,3}^+$, i.e. all elements of the form

$$c_0 + c_1 e_1 e_2 + c_2 e_1 e_3 + c_3 e_2 e_3, \quad c_0, c_1, c_2, c_3 \in \mathbb{R} \tag{6.59}$$

we can set $i = \mathbf{e}_1\mathbf{e}_2$ and $j = \mathbf{e}_2\mathbf{e}_3$ so that

$$\mathrm{tr}F_\Gamma f = \frac{1}{2}(I + S_\Gamma)f \tag{6.60}$$

$$= \frac{1}{2}f(\tilde{y}_1, \tilde{y}_2) + \frac{1}{2}\int_{\mathbb{R}^2} \frac{i(x_1 - \tilde{y}_1) + j(x_2 - \tilde{y}_2)}{|x - y|^2} f(x_1, x_2)\mathrm{d}x_1\mathrm{d}x_2. \tag{6.61}$$

Up to the factor 1/2 this is the monogenic signal $f_M = f + i\mathcal{R}_1 f + j\mathcal{R}_2 f := f + (i,j)\mathcal{R}f$ of Sommer and Felsberg [FS01]. More details about the Riesz transforms $\mathcal{R}_1$, $\mathcal{R}_2$ and $\mathcal{R}$ are given in the next section.

6.2.2.2 The Riesz transform and its property

We start by giving some general definitions and properties about the Riesz transform and see later how it can be understood as a particular extension of the Hilbert transform to higher dimensional spaces.

Riesz kernel

Definition 31 (Riesz transform). *The Riesz transform [Ste70] is defined as Fourier multipliers as:*

$$\widehat{\mathcal{R}f}(\xi) = \frac{i\xi}{\|\xi\|_2}\hat{f}(\xi) \tag{6.62}$$

$$\forall k \in \{1, \cdots n\}, \ \widehat{\mathcal{R}_k f}(\xi) = \frac{i\xi_k}{\|x\|_2}\hat{f}(\xi) \tag{6.63}$$

where $\|x\|_2 = \sqrt{\sum_i x_i^2}$.

or equivalently defined in the spatial domain by convolution with the n-dimensional Riesz kernel, for $k \in \{1, \cdots, n\}$

$$\mathcal{R}_k f = c_n \frac{x_k}{\|x\|_2^{n+1}} * f, \tag{6.64}$$

with c_n being a constant depending only on the dimension n, which can be computed by measuring the volume of the unit ball.

A unique generalization of the Hilbert transform Here we recall some characteristic properties of the Riesz transform. We see that under some rather general assumptions, the Riesz transform is the only valid candidate for higher dimensional generalizations of the Hilbert transform.

First, a translation operator will be denoted by $\tau_t, t \in \mathbb{R}^n$ and acts as follows $\tau_t f(x) = f(x - t)$. A dilation operator D_δ parameterized by $\delta > 0$ is defined as $D_\delta f(x) = f(\delta x)$. Finally, the set of all rotations parameterized by ρ, R_ρ, should also be considered. We may use equivalently its induced representation on function as $R_\rho f(x)$ or its effect on the running variables $f(\rho^{-1}x)$.

Property 12 (Characterization of the Hilbert transform). *Any operator T which is bounded on $L^2(\mathbb{R}^1)$ and fulfils the following set of conditions*

- *a) T commutes with translations,*

- *b) T commutes with positive dilations,*

6.2. GENERALIZATIONS TO HIGHER DIMENSIONS

c) T anti-commutes with the reflection $f(x) \mapsto f(-x)$,

is a constant multiple of the Hilbert transform.

Proof. We refer to the [Ste70] for more details.

We recall first the following lemma

Lemma 3 (Translation invariant function). *Let an operator T be bounded and linear from $L^2(\mathbb{R}^n)$ to itself, then T commutes with the translation if and only if there exists a bounded measurable function $m(\xi)$ so that $\widehat{Tf}(\xi) = m(\xi)\widehat{f}(\xi)$. And one has $\|T\| = \|m\|_\infty$*

So that due to the first condition, we have that it exists a bounded multiplier m such that $\widehat{Tf} = m\widehat{f}$. On the other side, the two last conditions may be rewritten as

$$TD_\delta = \text{sign}(\delta) D_\delta T \tag{6.65}$$

and basic Fourier calculus gives that dilations of factor δ of function f can be written in Fourier domain as

$$\widehat{D_\delta f} = |\delta|^{-1} D_{\delta^{-1}} \widehat{f} \tag{6.66}$$

So taking the Fourier transform of Eq. 6.65 yields, for all $f \in L^2$

$$D_\delta m \widehat{f} = D_\delta \mathcal{F}(Tf) = |\delta|^{-1} \mathcal{F}(D_{\delta^{-1}} Tf) = |\delta|^{-1} \text{sign}(\delta) \mathcal{F}\left(T(D_{\delta^{-1}} f)\right) \tag{6.67}$$

$$= \delta^{-1} m \widehat{D_{\delta^{-1}} f} = \delta^{-1} m |\delta| D_\delta \widehat{f} \tag{6.68}$$

$$D_\delta m \widehat{f} = \text{sign}(\delta) m D_\delta \widehat{f} \tag{6.69}$$

and the last equation should be valid everywhere in $\mathbb{R}$ which means that $\forall x \in \mathbb{R}, m(\delta x) = \text{sign}(\delta) m(x)$ which implies that m should be a constant multiplication of the sign function.

□

The idea now is to characterise functions with similar properties in higher dimensions. We will extend the coordinate wise operations.

Then basic Fourier calculus ensures

$$\widehat{\tau_t f}(\xi) = e^{i2\pi t \cdot \xi} \widehat{f}(\xi)$$

$$\widehat{D_\delta f}(\xi) = \delta^{-n} D_{\delta^{-1}} \widehat{f}(\xi)$$

$$\widehat{R_\rho f}(\xi) = R_\rho \widehat{f}(\xi)$$

If we moreover define $m(x) = (m_1(x), \cdots, m_n(x))$ as a set of n functions defined on $\mathbb{R}^n$, a rotation ρ can be written by means of its realization matrix $(\rho_{jk})_{j,k=1}^n$ (see for instance [Don11]), then for m being understood as a vector transformation, we have

$$m_j(\rho^{-1} x) = \sum_k \rho_{jk} m_k(x) \tag{6.70}$$

Property 13 (Characterization of the Riesz transforms). *Let $T_1, \cdots, T_n$ be n bounded operators on $\mathbb{R}^n$. If*

a) *For any $j \in \{1, \cdots, n\}$ and $t \in \mathbb{R}^n$, T_j commutes with τ_t,*

b) T_j commutes with all D_δ,

c) Given any rotation $\rho = (\rho_{jk})_{j,k=1}^n$, $\rho T_j \rho^{-1} f = \sum_k \rho_{jk} T_k f$,

then the T_j are constant multiples of the Riesz transforms.

Proof. The first point gives us, due to the lemma introduced in the previous property, that each bounded T_j can be written as a Fourier multiplier m_j. Moreover, the second point, using a same procedure as in the previous proof, for $\delta > 0$, yields $m_j(\delta x) = m(x)$ or, in other terms, all m_j are homogeneous of degree 0. We can apply the following lemma

Lemma 4. *If m is homogeneous of degree 0 such the following equality holds:*

$$m_j(\rho^{-1}x) = \sum_k \rho_{jk} m_k(x) \tag{6.71}$$

then m_j is a constant multiple of the jth Riesz transform:

$$m(x) = c \frac{x}{|x|} \tag{6.72}$$

Proof. Due to the normalisation, we consider the vectors x only on the unit sphere. Let $(e_1, \cdots, e_n)$ be the usual orthonormal basis and set $c = m_1(e_1)$ and therefore, $\forall j \neq 1, m_j(e_1) = 0$. Assume ρ keeps e_1 unchanged then we have, $\forall j \neq 1, m_j(e_1) = \sum_{k=2}^n \rho_{jk} m_k(e_1)$ which implies that $m_2(e_1) = \cdots = m_n(e_1) = 0$. So that, if we plug this last result in the condition of the lemma, we have $m_j(\rho^{-1}e_1) = \rho_{j1} m_1(e_1)$. And from x such that $\rho^{-1}e_1 = x$ it follows $\rho_{j1} = x_j$ which finishes the proof. □

and we get the desired conclusion of the theorem. □

The particular characterization gives a special property to the Riesz transform and also motivates its further study.

6.2.2.3 Signal analysis with the monogenic representation

As before, some local features of such monogenic signals can be computed. The three following features are once again described as *energetic, structural and geometric information* [FS01] and are illustrated, displayed on a sphere, in Fig. 6.3.

Local amplitude The local amplitude of a monogenic signal is defined in a similar way as for the analytic signal:

$$A_M(x,y) = \sqrt{|f(x,y)|^2 + |\mathcal{R}f(x,y)|^2} = \sqrt{f_M(x,y)\overline{f_M(x,y)}} \tag{6.73}$$

where the $\bar{\cdot}$ denotes the conjugation of a quaternion: for $q = q_0 + iq_1 + jq_2 + kq_3 \in \mathbb{H}$, $\bar{q} = q_0 - iq_1 - jq_2 - kq_3$.

Local phase

$$\varphi_M(x,y) = \arctan \frac{|\mathcal{R}f(x,y)|}{f(x,y)}, \tag{6.74}$$

6.3. ANALYSIS OF LOCAL MONOGENIC FEATURES

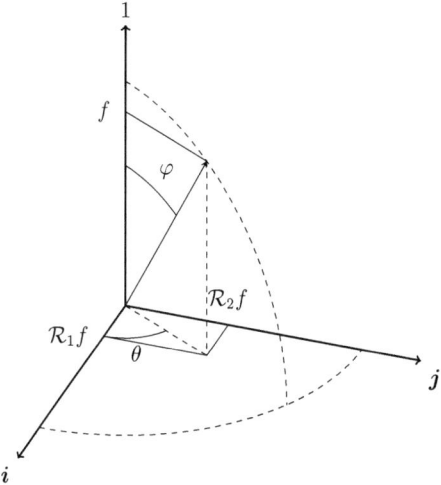

Figure 6.3: Spherical representation of the monogenic signal and its local features

and we still have that φ_M denotes the angle between f and f_M. This yields values $\varphi_M \in [-\pi/2; \pi/2]$. An alternative equivalent definition is using arccos:

$$\varphi_M = \arccos \frac{f}{|f_M|}. \qquad (6.75)$$

In this case, we have $\varphi_M \in [0; \pi]$

Local orientation Once again, we can derive an orientation $\theta_M \in [-\pi, \pi]$ based on the monogenic signal which represents the direction of the phase information.

$$\theta_M = \arctan \frac{\mathcal{R}_2 f}{\mathcal{R}_1 f} \qquad (6.76)$$

We note that this definition actually only provides an orientation mod. π. To determine the orientation resp. direction mod. 2π it needs a further orientation unwrapping step or sign estimation [LBO01, BPS00].

6.3 Analysis of local monogenic features

This section aims at explaining the usefulness of such local features for the analysis of image structural features. We first give some examples of computations to show the meanings of each local features in the context of *Amplitude Modulated-Frequency Modulated* (AM-FM) signals. We see in some cases, that the local features estimation is able to give an almost perfect description of both modulations.

Some more complicated cases are computed and show that one should be careful to respect the band-limited property which ensures the *split of identity* [FS01].

Finally, some ideas for image analysis based on such analytic representations are given based on local phase analysis.

6.3.1 Motivation

It is always admitted that before starting analyzing signals or images by means of analytic or monogenic representations, one should first proceed to several applications of band pass filters (as for instance explained in [FS01]).

However it might not be clear to the reader at the first glance, that this condition is of prime importance. To understand the problem, let's first have a look at some examples.

Amplitude and frequency cosine modulated Assume we are given a function f which can be written as a product of two functions which we call amplitude modulation $A(x)$ and frequency modulation $\varphi(x)$. It holds $f(x) = A(x)\cos(\varphi(x))$ where $x \in \mathbb{R}$. φ is called the phase or argument (this denomination becomes clear later). A is called the amplitude of the signal. It is also sometimes (abusively) denoted as envelope.

In this particular case, we consider an amplitude which is assumed to be a cosine wave. The phase φ depends here linearly on the space coordinate x. For some $0 < \omega_A < \omega_g$, we write $f(x) = \cos(\omega_A x)\cos(\omega_g x)$

We can rewrite, using trigonometric formulas $f(x) = A(x)g(x) = \frac{1}{2}(\cos((\omega_A + \omega_g)x) + \cos((\omega_g - \omega_A)x))$ and this leads to (using the Fourier transform of the distribution $\exp(ix)$, obtained by applying the phase shift theorem on the Fourier transform of the Delta distribution)

$$\widehat{f}(\xi) = \frac{1}{4}\left(\delta\left(\xi - (\omega_g + \omega_A)\right) + \delta\left(\xi + (\omega_g + \omega_A)\right)\right.$$
$$\left. + \delta\left(\xi - (\omega_g - \omega_A)\right) + \delta\left(\xi + (\omega_g - \omega_A)\right)\right), \tag{6.77}$$

$$\widehat{\mathcal{H}f}(\xi) = -\frac{i}{4}\left(\delta\left(\xi - (\omega_g + \omega_A)\right) - \delta\left(\xi + (\omega_g + \omega_A)\right)\right.$$
$$\left. + \delta\left(\xi - (\omega_g - \omega_A)\right) - \delta\left(\xi + (\omega_g - \omega_A)\right)\right), \tag{6.78}$$

$$\mathcal{H}f(x) = \frac{1}{2}\left(\sin\left((\omega_g + \omega_A)x\right) + \sin\left((\omega_g - \omega_A)x\right)\right), \tag{6.79}$$

$$f_A(x) = \frac{1}{2}e^{i(\omega_g + \omega_A)x} + \frac{1}{2}e^{i(\omega_g - \omega_A)x} = e^{i\omega_g x}\cos(\omega_A x). \tag{6.80}$$

Note that these computations are only valid under the assumption that $\omega_g - \omega_A > 0$ or equivalently $\omega_g > \omega_A$.

Based on the previous calculus we get the following set of local features

$$A_M(x) = |f_A(x)| = |\cos(\omega_A x)| = |A(x)| \tag{6.81}$$

$$\varphi(x) = \arg(f_A(x)) = \arctan\frac{\sin(\omega_g x)\cos(\omega_A x)}{\cos(\omega_g x)\cos(\omega_A x)} = \omega_g x \bmod 2\pi \tag{6.82}$$

In this case one should notice that the phase depends only on the frequency modulations while the local amplitudes are uniquely characterised by the amplitude modulations, appearing as the low frequency component. This is exactly the idea behind the split of identity: under certain frequency conditions (mainly that the signal should be band limited) the decomposition into local amplitude and local phase is orthogonal, in the sense that changing one feature does not interfere with the other feature.

6.3. ANALYSIS OF LOCAL MONOGENIC FEATURES

Note moreover that this case is a typical case of application of the Bedrosian identity which reads [Bed63]

Theorem 14 (Bedrosian identity). *If f and g are two complex analytic functions of the real variable t whose Fourier spectrum are non-overlapping (i.e. $\exists a \in \mathbb{R}^+ : \mathcal{F}(f)(\xi) = 0, \forall |\xi| < a$ and $\mathcal{F}(g)(\xi) = 0, \forall |\xi| > a$) then we have:*

$$\mathcal{H}(fg) = f\mathcal{H}(g) \tag{6.83}$$

For stability reasons, the amplitude modulations are more often given as $A(x) = 1 + \frac{1}{2}\cos(\omega_A x)$ and $f(x) = \cos(\omega_g x) + \frac{1}{2}\cos(\omega_A x)\cos(\omega_g x)$. Based on the previous calculus we get

$$\mathcal{H}f(x) = \mathcal{H}\cos(\omega_g \cdot)(x) + \frac{1}{2}\mathcal{H}\cos(\omega_g \cdot)\cos(\omega_A \cdot)(x)$$
$$= \sin(\omega_g x) + \frac{1}{2}\cos(\omega_A x)\sin(\omega_g x)$$
$$= A(x)\sin(\omega_g x)$$
$$f_A(x) = A(x)e^{i\omega_g x}$$

From what it follows

$$A_A(x) = A(x)$$
$$\tan(\varphi(x)) = \tan(\omega_g x)$$

once again, the local amplitude is determined by the amplitude $A(x)$ and the local frequency only depends on the frequency modulations component.

Dirac impulse Here we have a look at another example to show that this split of identity is actually a particular property. If we consider our input signal to be a Dirac impulse *i.e.* $f(x) = \delta(x)$ we can compute the Hilbert transform of it (this time in the time domain, as the Dirac distribution is the identity of the convolution) and we get

$$\mathcal{H}f(x) = -\frac{1}{\pi x} \tag{6.84}$$

$$f_A(x) = \delta(x) - \frac{i}{\pi x} \tag{6.85}$$

which leads to the following local features:

$$A_A(x) = |f_A(x)| = \frac{1}{\pi |x|} \tag{6.86}$$

$$\varphi(x) = -\operatorname{sign}(x)\frac{\pi}{2} \tag{6.87}$$

In this case, the signal is definitely not band limited as the Fourier transform of the Dirac distribution is the constant distribution equals to 1. As we can see, even if the amplitude of the signal is 0 almost every where in $\mathbb{R}$, this is not the case for the local amplitude as the Hilbert transform convey the singularity (due to the convolution) at 0 along the space or time axis.

6.3.2 The not trivial case

In this section we start by investigating the importance of having a band-limited signal in order to get a proper split of identity. Ideally we would like to have estimates about how robust and reliable the features are in order to describe the signals' local properties. While this is still a work in progress, we however give some ideas and detail some calculations on how it could be handled.

Assume the amplitude can be described by a sum of cosine waves at different (positive) frequency in Ω_A. Assume moreover that the frequency modulations can be described by a sum of cosine waves too, with frequencies in Ω_g. In addition assume that the frequencies oscillate faster than any of the amplitude modulation, which leads to $\forall \omega_A \in \Omega_A, \omega_g \in \Omega_g, \omega_A < \omega_g$

Following these assumptions we have the following signal

$$f(x) = \sum_{\omega_A \in \Omega_A} \cos(\omega_A x) \sum_{\omega_g \in \Omega_g} \cos(\omega_g x),$$

which can be written as

$$\sum_{\omega_A \in \Omega_A, \omega_g \in \Omega_G} \cos(\omega_A x)\cos(\omega_g x),$$

so that we actually have a linear combination of different functions similar to the signal in the previous signal, and we get directly

$$f(x) = \left(\sum_{\omega_A \in \Omega_A} \cos(\omega_A)\right)\left(\sum_{\omega_g \in \Omega_g} \cos(\omega_g)\right) \quad (6.88)$$

$$\mathcal{H}f(x) = \left(\sum_{\omega_A \in \Omega_A} \cos(\omega_A)\right)\left(\sum_{\omega_g \in \Omega_g} \sin(\omega_g)\right) \quad (6.89)$$

$$f_A(x) = \sum_{\omega_A \in \Omega_A, \omega_g \in \Omega_g} e^{i\omega_g x} \cos(\omega_A x) \quad (6.90)$$

We can finally compute the following local features:

$$A_A^2(x) = \left(\sum_{\omega_A \in \Omega_A} \cos(\omega_A)\right)^2 \left(\left(\sum_{\omega_g \in \Omega_g} \cos(\omega_g)\right)^2 + \left(\sum_{\omega_g \in \Omega_g} \sin(\omega_g)\right)^2\right) \quad (6.91)$$

$$\tan(\varphi_A(x)) = \frac{\sum_{\omega_g \in \Omega_g} \sin(\omega_g x)}{\sum_{\omega_g \in \Omega_g} \cos(\omega_g x)} \quad (6.92)$$

Now it appears that the frequency modulations do have an effect on the local amplitude, even if the local frequency is only characterized by the original frequency modulations.

We want now to analyse what is the effect of having more spectral components. As a first example, we consider the case where the frequency modulations are symmetric around a certain reference frequency ω_{g0} *i.e.*

$$\exists \{\delta_g\}_g : \Omega_g := \bigcup_{\delta_g} \{\omega_{g0} + \delta_g, \omega_{g0} - \delta_g\} \quad (6.93)$$

6.3. ANALYSIS OF LOCAL MONOGENIC FEATURES

so that the phase can now be written as

$$\tan(\varphi(x)) = \frac{\sum_{\delta_g} \sin(\omega_{g_0} x + \delta_g x) + \sin(\omega_{g_0} x - \delta_g x)}{\sum_{\delta_g} \cos(\omega_{g_0} x + \delta_g x) + \cos(\omega_{g_0} x - \delta_g x)} \tag{6.94}$$

$$= \frac{\sum_{\delta_g} 2 \sin(\omega_{g_0} x) \cos(\delta_g)}{\sum_{\delta_g} 2 \cos(\omega_{g_0} x) \cos(\delta_g)} \tag{6.95}$$

$$= \tan(\omega_{g_0} x) \frac{\sum_{\delta_g} \cos(\delta_g)}{\sum_{\delta_g} \cos(\delta_g)} \tag{6.96}$$

$$= \tan(\omega_{g_0} x) \tag{6.97}$$

even if the different frequency components have different weights of importance. The only condition for this calculation to hold is that the frequency amplitudes are distributed symmetrically centered at ω_{g_0}. Indeed, if we consider the case of symmetric frequencies with symmetric weights, they will cancel out in the calculation of $\tan(\varphi(x))$ as in the previous case.

Altogether it means that whenever the frequency modulations are symmetric centered at a given ω_{g_0}, the local phase can be perfectly computed, knowing only the centre of the frequency components.

Be aware that all these calculations are only valid when the domains of both amplitude and frequency modulations' spectra in the Fourier domain are non-overlapping.

The local amplitude is a little more complicated to deal with.

$$A_A(x)^2 = \left(\sum_{\omega_A} \cos(\omega_A x)\right)^2 \left(\left(\sum_{\omega_g} \cos(\omega_g x)\right)^2 + \left(\sum_{\omega_g} \sin(\omega_g x)\right)^2\right)$$

$$= \left(\sum_{\omega_A} \cos(\omega_A x)\right)^2 \left(\sum_{\omega_g} \cos(\omega_g x)^2 + \sum_{\omega_g} \sin(\omega_g x)^2 + 2 \sum_{\substack{\omega_g, \omega'_g \\ \omega_g \neq \omega'_g}} \cos(\omega_g x) \cos(\omega'_g x) + \sin(\omega_g x) \sin(\omega'_g x)\right)$$

$$= \left(\sum_{\omega_A} \cos(\omega_A x)\right)^2 \left(|\Omega_g| + 2 \sum_{\substack{\omega_g, \omega'_g \\ \omega_g \neq \omega'_g}} \cos(\omega_g x) \cos(\omega'_g x) + \sin(\omega_g x) \sin(\omega'_g x)\right)$$

$$= \left(\sum_{\omega_A} \cos(\omega_A x)\right)^2 \left(|\Omega_g| + 2 \sum_{\substack{\omega_g, \omega'_g \\ \omega_g \neq \omega'_g}} \cos\left((\omega_g - \omega'_g) x\right)\right)$$

$$= \left(\sum_{\omega_A} \cos(\omega_A x)\right)^2 \left(|\Omega_g| + 2 \sum_{\substack{\delta_g, \delta'_g \\ \delta_g \neq \delta'_g}} \cos\left((\delta_g - \delta'_g) x\right)\right)$$

$$A_A(x) \approx \left|\sum_{\omega_A \in \Omega_A} \cos(\omega_A x)\right| \cdot \sqrt{2} |\Omega_g|$$

These last calculations are only valid for small δ_g. And in this case, the local amplitude scales linearly with the number of waves in the frequency modulations. Note furthermore that ensuring a bandlimited amplitude modulation is not sufficient to get a split of identity, however in the case of many amplitude modulation components, the signal no longer has an intrinsic dimensionality of one. This whole framework however yields only one potential candidate for the local amplitude and local frequency. In other terms, the last estimation computed actually combines all components of all wave modulation, as they all appear in the sum and return some kind of average.

6.3.3 The higher dimensional case

Intrinsic one dimensional image We start our examples by looking at a two dimensional function which has an intrinsic dimensionality of 1. It is similar to the examples we used in the chapter about the two dimensional discrepancy approximation and is depicted on Fig. 4.6(b). Let $f(x_1, x_2) = \cos(\alpha_1 x_1 + \alpha_2 x_2)$ which we can also write as $f(x) = \cos(\alpha^T x)$ where $x = (x_1, x_2)$ corresponds to the spatial coordinates. As we did in the one dimensional case, we can compute the Fourier transform of the Dirac distribution which yields $\hat{\delta} = 1$ and by applying the Fourier shift theorem, we get that

$$\hat{f}(\xi) = \delta_\alpha(\xi) + \delta_{-\alpha}(\xi), \tag{6.98}$$

where we have defined $\delta_\alpha(\xi) := \tau_\alpha \delta(\xi) = \delta(\xi - \alpha)$. If we finally plug in this result in the monogenic framework we get

$$\widehat{\mathcal{R}_1 f}(\xi) = i\frac{\xi_1}{\|\xi\|}\frac{1}{2}(\delta_\alpha(\xi) + \delta_{-\alpha}(\xi)) \tag{6.99}$$

$$= \frac{i}{2}\frac{-\alpha_1}{\|\alpha\|} + \frac{i}{2}\frac{\alpha_1}{\|\alpha\|} \tag{6.100}$$

$$= \frac{i}{2}\frac{\alpha_1}{\|\alpha\|}(\delta_\alpha(\xi) - \delta_{-\alpha}(\xi)) \tag{6.101}$$

$$\mathcal{R}_1 f(x) = \frac{\alpha_1}{\|\alpha\|}\sin(x^T \alpha) \tag{6.102}$$

$$\mathcal{R}_2 f(x) = \frac{\alpha_2}{\|\alpha\|}\sin(x^T \alpha) \tag{6.103}$$

$$f_M(x) = \cos(x^T \alpha) + \frac{i\alpha_1 \sin(x^T \alpha) + j\alpha_1 \sin(x^T \alpha)}{\|\alpha\|} \tag{6.104}$$

In terms of local features, we have

$$A_M(x) = \left(\cos(x^T\alpha)^2 + \frac{\alpha_1^2 + \alpha_2^2}{\|\alpha\|^2}\sin(x^T\alpha)^2\right)^{1/2} = 1, \tag{6.105}$$

$$\varphi_M(x) = \arctan\left(\frac{\frac{\alpha_1^2+\alpha_2^2}{\|\alpha\|^2}\sin(x^T\alpha)^2}{\cos(x^T\alpha)}\right) = x^T\alpha\,\mathrm{sign}\left(\sin(x^T\alpha)\right), \tag{6.106}$$

$$\theta_M(x) = \arctan\frac{\frac{\alpha_2}{\|\alpha\|}\sin(x^T\alpha)}{\frac{\alpha_1}{\|\alpha\|}\sin(x^T\alpha)} = \arctan\frac{\alpha_2}{\alpha_1}, \tag{6.107}$$

and this calculation coincide with the results expressed by Unser et al. [USVDV09] regarding the combination of monogenic representations, Hilbert transforms and Radon transforms.

We can notice here again, as in the one dimensional case, that the features are completely independent from one another. While this example is rather simple, similar conclusions can be drawn when considering more complex amplitude modulations (with some conditions).

Dirac impulse We consider in this case a function f expressed as a Dirac centered at a certain point $p = (p_1, p_2) \in \mathbb{R}^2$: $f(x) = \delta_p(x) = \delta(x - p)$

A direct calculation (using the fact that the Dirac impulse is the unity for the convolution operation) yields

$$\mathcal{R}_1 f(x) = c_2 \frac{x_1 - p_1}{\|x - p\|^3} \qquad (6.108)$$

$$\mathcal{R}_2 f(x) = c_2 \frac{x_2 - p_2}{\|x - p\|^3} \qquad (6.109)$$

$$f_M(x) = \delta(x) + c_2 \frac{i(x_1 - p_1) + j(x_2 - p_2)}{\|x - p\|^3} \qquad (6.110)$$

which means, in terms of local features

$$A_M(x) = \left(\delta_p(x)^2 + c^2 \frac{(x_1 - p_1)^2 + (x_2 - p_2)^2}{\|x - p\|^6}\right)^{1/2} = \begin{cases} 1 & x = p \\ \frac{c}{\|x\|^2} & x \neq p \end{cases} \qquad (6.111)$$

$$\varphi_M(x) = \begin{cases} 0 & x = p \\ \frac{\pi}{2} & x \neq p \end{cases} \qquad (6.112)$$

$$\theta_M(x) = \arctan \frac{x_2 - p_2}{x_1 - p_1} \qquad (6.113)$$

In this case, the split-of-identity is definitely not respected due to the infinite extension of the Fourier spectrum. Moreover, we see that the phase does not really yield any valuable information. The orientation on the other hand gives the position of the current point x relatively to the given centre of the Dirac impulse p.

6.3.4 Interpretation of phase and amplitude

6.3.4.1 Qualitative results

We give here an idea about how the local features can help characterising the local structures of an image.

Here, we interpret, and we give more details about this in the next section, the Hilbert transform as a normalized derivative or both Riesz transforms as first and second normalized partial derivatives along the coordinate axis. With this new insight we can say that $\tan(\varphi)$ is the ratio of changes in the image over the intensity information. This means that a value close to 0 can appear only when the Hilbert transform is small compared to the original function; which in turns means that there are only a few changes relative to the signal's intensity. Therefore we can describe this region as a rather uniform one.

Conversely, if we have $|\tan(\varphi)| \to \infty$ we can assume that the changes are important compared to the original image which turns out to be considered as abrupt changes in the signal, or in other words, when a structure is appearing.

Similarly, for the higher dimensional case, the Riesz transforms act as normalised partial derivatives and we can carry out the same analysis. $\tan(\varphi_M)$ gives us a ratio of change over an intensity information. If this value is close to 0 we can consider the local neighborhood as rather uniform while an increasing value tends to show stronger variations in at least one direction. The preferred direction can be computed using the orientation information, which is the ratio of the second derivative over the first one.

In both one and two dimensional cases, the amplitude acts as an estimate of the importance of the local structure computed. It indeed tells us whether or not one at least of the components has a strong impact. As an example, we can consider the case that $f(x)$ and $\mathcal{H}f(x)$ are small then even if the local phase may tend to ∞ (*i.e.* strong changes instead of uniformity), the local amplitude (close to 0) tells us that the local features are not reliable.

6.3.4.2 Quantitative analysis

Phase mutual information This image analysis approach has been developed by Mellor and Brady [MB04, MB05] as an improvement (in certain cases) to the classical intensity mutual information [WIVA+96]. The idea underneath using local phase estimations instead of raw intensity information is that mutual information assumes a meaningful relationship between the data compared. Intensities are unfortunately not robust against contrast or illumination changes. On the opposite, local phases (and therefore analytic representations) give insights about the structures in an image. Therefore it yields some kind of invariance to contrast or illumination.

The authors investigate in particular the problem of medical image alignment and fusion. When images are coming from, say ultrasound images, and need to be registered onto other types of images, say magnetic resonance images, the dynamics as well as the intensities' distribution are quite different. However, the structures are preserved from one image to the other. This is due to the fact that phase are responsible for the structural information while the local energy serves as an estimation of the local enveloppe.

Phase congruency Phase congruency is a concept introduced in its current form, to our knowledge, in the mid 90s. It is said to overcome many classical phase-based methods in recognizing structures due to the fact that it gives a global, data-independent, metric. It varies between -1 and 1 independently from the lightning conditions. This simplifies algorithms looking for certain thresholds in order to decide whether an edge is present at a given pixel or not. It has first been applied, among others, to detect symmetries and asymmetries in intensities [Kov97].

The phase congruency in the one dimensional case is defined as [Kov99]

$$PC(x) := \max_{\overline{\phi} \in [0, 2\pi)} \frac{\sum_k A_k \cos\left(\phi_k(x) - \overline{\phi}(x)\right)}{\sum_k A_k}, \tag{6.114}$$

with A_k representing the amplitude of the of the kth Fourier coefficient and it can be shown [VO89] that this quantity is proportional to the local amplitude of the analytic signal:

$$A(x) = PC(x) \sum_k A_k.$$

Unfortunately, this metric is not invariant to illumination changes. In its work, Kovesi computed the different frequency components using log-Gabor wavelets as suggested by the visual system [Fie87]. This scale decomposition is done using n quadrature filters and frequencies and amplitudes are com-

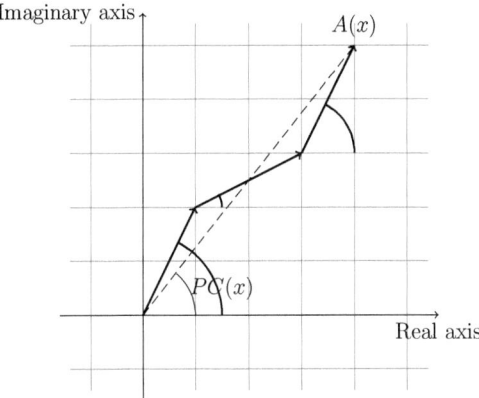

Figure 6.4: A graphical representation of the phase congruency as suggested by Peter Kovesi [Kov99]. The blue arc corresponds to the phase congruency while the black ones corresponds to the phase of different log-Gabor wavelets.

puted at each scales and used as input of the phase congruency calculation. Further studies regarding its applicability with regard to edges and corner detection are analysed [Kov03, Kov02]

6.4 Comparing higher dimensional analytic signals

We want here to illustrate the differences between the generalizations proposed. We visually assess the characteristics of both approaches first facing a Siemens star[3] then facing a checkerboard image. Both examples are interesting for their regularity (point symmetry for the star and many horizontal and vertical line symmetries for the checkerboard).

An example of such star is depicted on Fig. 6.5(a). The two other images of first row from Fig. 6.5 illustrate the two components of the Riesz transform. As we can see, and we will come back on that property later, the partial Riesz transforms show in some ways a similar behavior as steered derivatives. The first component tends to emphasize horizontal edges while the second one tends to respond to vertical ones.

The second row shows the results applying the different Hilbert transforms to the Siemens star. The two first images represent the results of the two partial Hilbert transforms and the last one depicts the results after the total Hilbert transform. We can notice the high anisotropy of these transforms at, for instance, the strong vertical resp. horizontal delimitation through the centres of the images. We can also notice the patchy responses of the total Hilbert transform.

As the Riesz kernel in polar coordinate $[r, \alpha]$ of the spatial domain reads

$$R(r, \alpha) \sim \frac{1}{r^2} e^{i\alpha} \qquad (6.115)$$

---

[3]The Siemens star is a known test image to characterize the resolution of different optical/graphical devices such as printers or projectors. It is interesting as it shows lots of regularity, many intrinsic one dimensional and two dimensional parts.

it exhibits an isotropic behavior with respect to its magnitude. In comparison, the partial and the total Hilbert transforms induce a strict relationship to the orthogonal coordinate system and therefore also the two-dimensional analytic signal is coined in such a way.

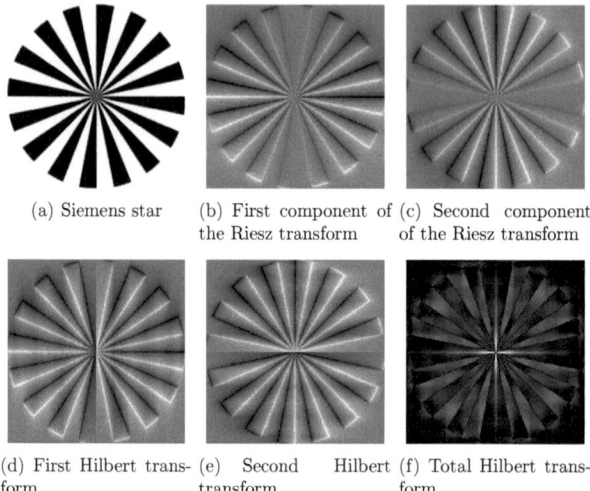

(a) Siemens star   (b) First component of the Riesz transform   (c) Second component of the Riesz transform

(d) First Hilbert transform   (e) Second Hilbert transform   (f) Total Hilbert transform

Figure 6.5: The Siemens star together with the different Riesz and Hilbert transforms presented in this section.

Next we consider the local features computed according to the formulas introduced above. The results are depicted in Fig. 6.6. The first row corresponds to monogenic features, while the second one corresponds to analytic features[4]. The last column shows the orientations whose intensity is weighted proportionally to the cosine of the phase: strength (cosine of the phase) is encoded as an intensity value of the color and the color itself corresponds to the orientation. The main differences between these two sets of features lie in the shape or boundaries. While monogenic features yield rather smooth boundaries, the analytic representation creates abrupt changes due to its anisotropy. We can remark how the phase gives reasonable insights about the structures in the images.

In comparison to the Siemens star, the checkerboard example (see Fig. 6.7(a)) shows many orthogonal features. In this case, we see that the partial Hilbert transforms give some good insights of the closeness of an edge and preserves the checkerboard structure (Fig. 6.7(d) and 6.7(e)) while the Riesz transform gives more local responses. The total Hilbert transform acts as an accurate corner detection, as it can be seen from its response on Fig. 6.7(f).

When discussing the analytic and monogenic features (Fig. 6.8) we remark that this effect is preserved. The Riesz transform being well localized at the edges does not yield many differences inside one of the squares and seems to jump from an extreme to another through those edges. See in particular Fig. 6.8(b) for an illustrative example of the phase. On the other side, the Hilbert transform containing more neighborhood information yields smoother transition in the phase from a square to another. This idea has to be considered carefully based on the applications one is interested in.

---

[4]To simplify the publication process, the figures are displayed in gray level. Color images may be resquested from the author - this holds for any figure in this document.

6.4. COMPARING HIGHER DIMENSIONAL ANALYTIC SIGNALS

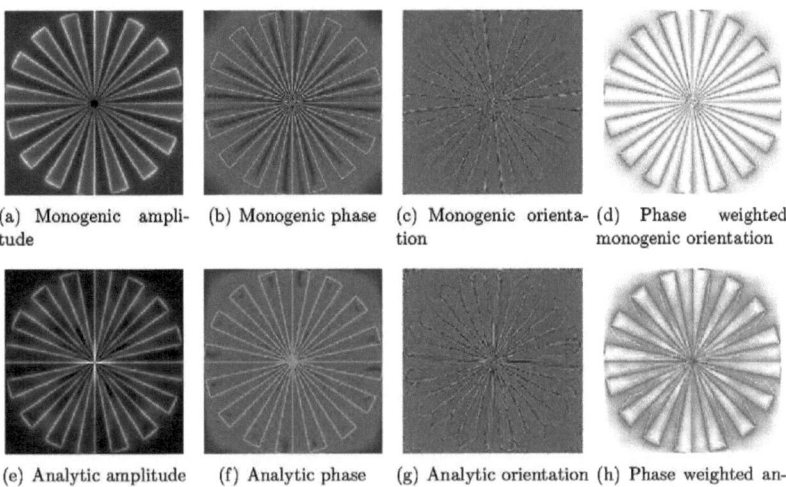

(a) Monogenic amplitude (b) Monogenic phase (c) Monogenic orientation (d) Phase weighted monogenic orientation

(e) Analytic amplitude (f) Analytic phase (g) Analytic orientation (h) Phase weighted analytic orientation

Figure 6.6: Local features computed with the monogenic signal representation (first row) and the multidimensional analytic signal (second row).

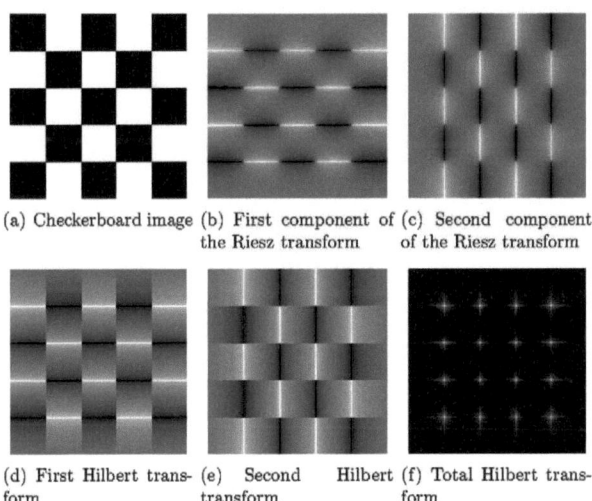

(a) Checkerboard image (b) First component of the Riesz transform (c) Second component of the Riesz transform

(d) First Hilbert transform (e) Second Hilbert transform (f) Total Hilbert transform

Figure 6.7: The checkerboard together with the different Riesz and Hilbert transforms presented in this section.

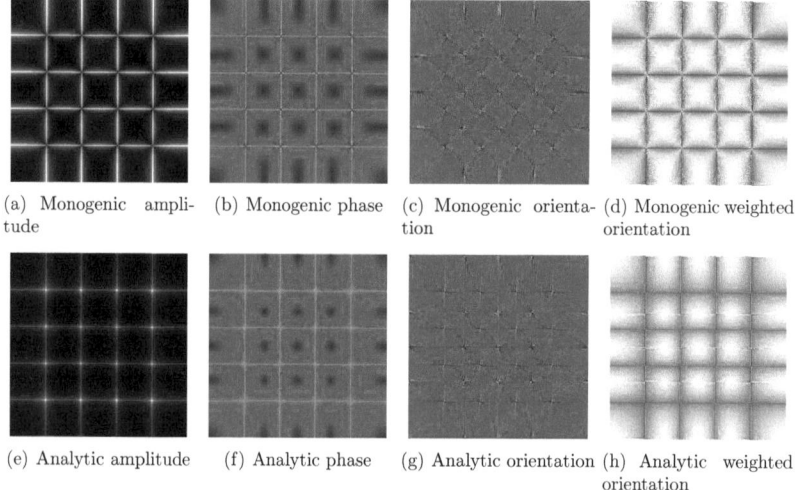

(a) Monogenic amplitude (b) Monogenic phase (c) Monogenic orientation (d) Monogenic weighted orientation

(e) Analytic amplitude (f) Analytic phase (g) Analytic orientation (h) Analytic weighted orientation

Figure 6.8: Local features computed with the monogenic signal representation (first row) and the multidimensional analytic signal (second row).

Part IV

Applications

Chapter 7

Defect detection in regularly textured patterns

We study in this chapter a novel algorithm used for defect detection in regularly textured surfaces. It is divided into three sections the first one describes the general context. The second one introduces and develops our algorithm first presented in [BSM11] and extended and improved in [SBHM12]. Finally some examples and comparison with other methods are given.

This chapter is not intended as a detailed state-of-the-art in defect detection but we refer the reader to the above given references for a rather exhaustive study (or directly look at [Xie08, Kum08]).

7.1 Problems and Restrictions

7.1.1 Problem statement

We want here to solve the rather general problem:

Given an image flow of a fabric, can we tell if there are any defects in the finished product? If yes the location of the defect should be given.

There are two points to work on here: the first one is to find out whether an image contains a defect or not. This is known as a *detection* problem in the computer vision literature. The other point in the sentence is to locate this particular defect. We call it a *localization* problem. We could eventually add a third task to the two others which would be to recognise the type of defect; which would be a classification problem. For some applications, a (even big) scratch might have a smaller importance than a small hole. In such case, different decisions can be taken: should we send the product anyway? or should we remanufacture it? Such considerations have to be kept in mind for practical industrial applications.

Here we want to keep our algorithm independent from the type of defect and leave this problem to a human expert. This is already a major difference with some other researchers who try to find a particular shape of defect. See for instance the work of Song *et al.* [SPK95] where the authors try to detect cracks and only cracks. Such systems even though they are really reliable for a certain type of defect, lack of generality and cumbersome tunings need to be done for them to work on more general defects.

7.1.2 Restrictions

As we detail in the next section, we introduce the discrepancy norm as an objective function for a template matching approach. As we have already seen, this (dis-)similarity measure is particularly suited for measuring misalignments between patterns. We therefore restrict our approach to textures showing a strong periodicity. Therefore we can ensure a rather wide area of convergence compared to other similarities. The idea is that thanks to the monotonicity property of the misalignment function, we can use the discrepancy norm with some local optimisation procedure to ensure a good, if not perfect, alignment with a golden pattern. If this alignment is not good enough, we can consider the location being analyzed as a defect, or at least containing one.

7.1.3 Motivations

As an example and a motivation to justify the use of pixel-based similarity measures for template matching, let us have a look at some particular textures. These textures contain different kind of periodicity and different kind of defects. We see that using a SIFT [Low04] feature extraction can not be used for this kind of defect detection task.

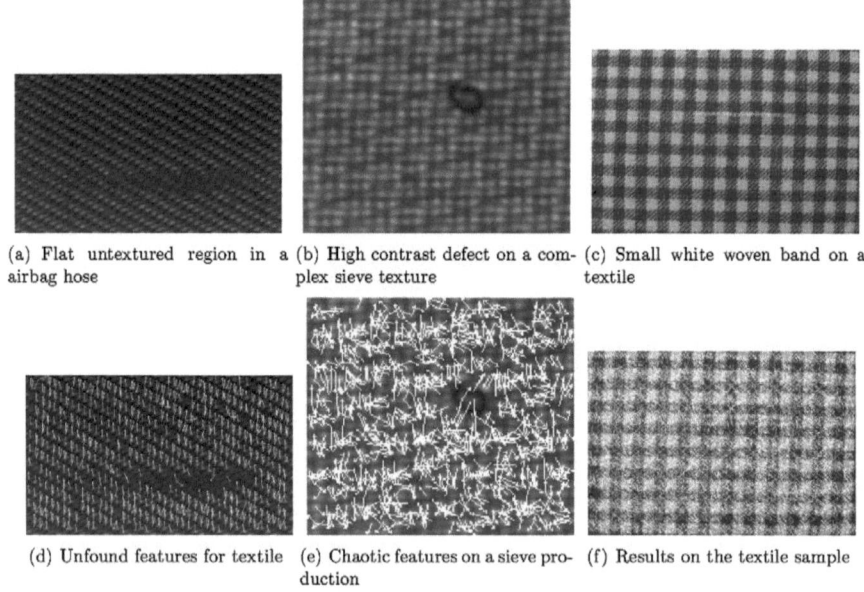

(a) Flat untextured region in a airbag hose (b) High contrast defect on a complex sieve texture (c) Small white woven band on a textile

(d) Unfound features for textile (e) Chaotic features on a sieve production (f) Results on the textile sample

Figure 7.1: Some examples of SIFT keypoints detection on defective image of different textures. The first row shows the gray-level images while the second row shows the corresponding SIFT keypoints found.

The first sample is an example of a defect represented as a uniform untextured region (see Fig. 7.1(a)). In this case, the SIFT keypoint detector fails at finding any features in this region (as seen on Fig. 7.1(d)). Therefore this defect would definitely not be found in a classifier.

The second sample considers a dense sieve texture with low contrast and rather high complexity corrupted by a black circle (Fig. 7.1(b)). Here, as we can see on Fig. 7.1(e), the features found are rather chaotic. Therefore a classifier, such as an SVM classifier, would have problem finding the boundaries (in the kernel or feature space) between good samples and defect ones. Moreover, only few features appear on the black circle.

Finally, the last example is taken from the TILDA database and represents a textile. This sample is corrupted by a small white band on the lower part of the image (see Fig. 7.1(c)). On the result image, Fig. 7.1(f), we see two main kinds of features which are directed along the two diagonals. However, the defect seems to be completely nonexistent in the SIFT keypoint representation.

We could however note that using keypoints descriptor on a dense grid, such as HOG [DT05] features would get rid of the problems regarding the two first samples (non detection of keypoints). However, even such a keypoint description would not have any positive outcome on the last sample.

7.2 A Template-Matching Algorithm

Let us now have a closer look to our algorithm. While many works have been done on feature extraction [BBL02] or statistical analysis of textural characteristics [HSD73] only few effort has be done on template matching approaches. We claim that this is mainly due to the the weakness of known similarities for such tasks (as we see in Fig.7.2). This is why we investigate this idea using the discrepancy norm as an objective function.

7.2.1 The original idea

In our first approach we give a brute force algorithm which is based on locally assessing the quality of a texture compared to a golden template. To this purpose we are given an $M \times N$ reference pattern R which we know is defect-free and covers at least (but more is better) one full period of the texture. Then an input image I of size $L \times W$ is being analysed with a sliding window approach. Assume the texture has a period of m pixels in one direction and n in the other, for each location (x, y) with $x \leq L$ and $y \leq W$ from the input image I a small patch $A_{x,y}$ of size $m \times n$ centered at pixel (x, y) is taken and being compared to all possible location of the reference. So we actually have two nested sliding windows: one along the input image and one for each patch of the input image in the reference template. This is not efficient enough for industrial applications but it fixes the first brick for a more evolved algorithm. This dummy brute-force approach is summarised in algorithm 1

Fig. 7.2 shows some examples of positive response of our algorithm to defects when either using the discrepancy norm (last column) or the Euclidean norm (second column). As we were expecting, the L^2 norm does not manage to discriminate properly defective region from defect free ones

7.2.2 Improvements

As already stated above, this approach is not tractable for industrial purposes due (mainly) to the two nested sliding windows. The first improvement one can think of is to make use of the monotonicity of the discrepancy norm when facing misaligned functions. This allows us to get rid of the time consuming sliding window on the size of the reference template. This optimisation can be done by means of local semi-Newton optimisation, for instance with BFGS update proposed by Broyden [Bro70a, Bro70b] ,

Algorithm 1 Local Defectiveness Map

1: **function** LDMAP(referenceImage, defectImage(size $M \times N$))
2:    extract $m \times n$ patch from referenceImage, where $m \times n$ covers 1-2 periods of pattern
3:    **for** $x = 0, n, 2*n, \cdots \div (N-n)$ **do**
4:      **for** $y = 0, m, 2*m, \cdots \div (M-m)$ **do**
5:         set position of $m \times n$ sliding window to (x, y) of defectImage
6:         **for** all pixels $p = (i, j)$ of sliding window **do**
7:            center reference patch on pixel p
8:            calculate discrepancy between sliding window and reference patch
9:            enter dissimilarity value in (x, y) of resultImage
10:         **end for**
11:      **end for**
12:    **end for**
13:    aggregate temporary images with min-function in an area of δ pixels
14:    binarize resultImage
15:    **return** resultImage
16: **end function**.

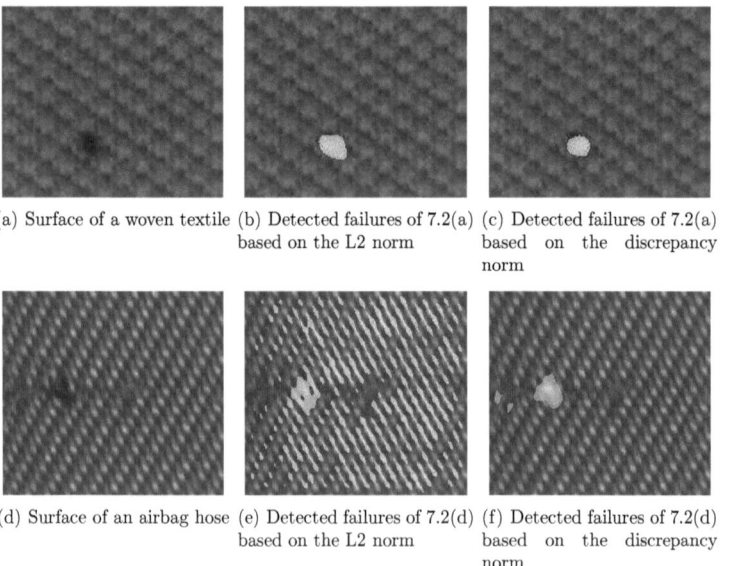

(a) Surface of a woven textile
(b) Detected failures of 7.2(a) based on the L2 norm
(c) Detected failures of 7.2(a) based on the discrepancy norm
(d) Surface of an airbag hose
(e) Detected failures of 7.2(d) based on the L2 norm
(f) Detected failures of 7.2(d) based on the discrepancy norm

Figure 7.2: Defect detection in textile and airbag hose production after a threshold on the computed LDMap with either the L_2 or the discrepancy norm. Threshold of DN images is 0.8 and of L_2 0.7, with normalized values between 0 and 1.

Fletcher [Fle70], Goldfarb [Gol70] and Shanno [Sha70]. However, due to the local property of such optimisation a smart first guess has to be given to the algorithm for better results.

In a first time, as used in [BSM11] `Direct` [JPS93, Shu72] iterations had been added as global optimisation procedure. This optimisation algorithm is particularly suited for Lipschitz functions and therefore got our interest.

However even if this is a deterministic optimisation procedure, its convergence can be slow and depends strongly on the current data. Therefore one of the major improvement added in [SBHM12] is by taking a first preprocessing step based on the RANSAC algorithm [FB81]. This allows cumbersome comparisons with many different patches due to the `Direct` algorithm and replace them by some chosen candidates. Then the outcome of the first RANSAC comparison is used for further local optimisation as before.

A final improvement, already present in the earlier version of the algorithm is done by applying a coarse-to-fine search. This framework allows us to quickly label large defect free region as so. Whenever the algorithm is not sure whether a defect is present or not, it refines at a finer scale. This yields also the patchy results as depicted in the next section. Furthermore, instead of having a sliding window on the input image too, we split the image into different patches with a certain overlap (in our experiments we have chosen to have an order of 10% of the patch size overlapping), this improves the results, when the defect is exactly located at a border of a patch.

Finally as it has been noted earlier, we need to define a size for the periodicity in order to define our patch size. This problem has been resolved in our latest algorithm by analysing the homogeneities of different patches at different sizes [LMSYAR11]. This approach seems to be both robust and efficient.

7.3 Experiments - Results

Some further experiments have been described in [SBHM12] and we only give a short description for illustration purposes.

We follow here the idea suggested in Tolba *et al.* [TKMA10] and use the following criteria for the performance evaluation

$$PCD = (1 - (FAR + FRR)) \times 100, \tag{7.1}$$

where FAR and FRR denote the False Alarm Rate and the False Rejection Rate respectively. Table 7.1 summarizes the overall performance with state of the art algorithms.

As we can see our approach is not yet competitive with state of the art methods but it introduces a novel concept (namely the template matching approach) in the context of automatic defect detection on textured surfaces. Further research has to be done to increase the overall results. Some results of our algorithms are depicted in Fig. 7.3

These images show the robustness of the approach against different kind of images. For instance image 7.3(h) shows only small contrast between defective regions and non defective ones.

As a last argument we show some illustrative examples of our method compared to a One-Class SVM framework (OC-SVM) used as a novelty detector[SWS$^+$00], as suggested in [JBPN09, TKMB09]. Here we have considered a simple OC-SVM learned directly on the grey-level intensities without any other preprocessing but an histogram equalization. The kernel used here is a Gaussian kernel.

The OC-SVM framework allows an automatic clustering of some given template. It tries to separate

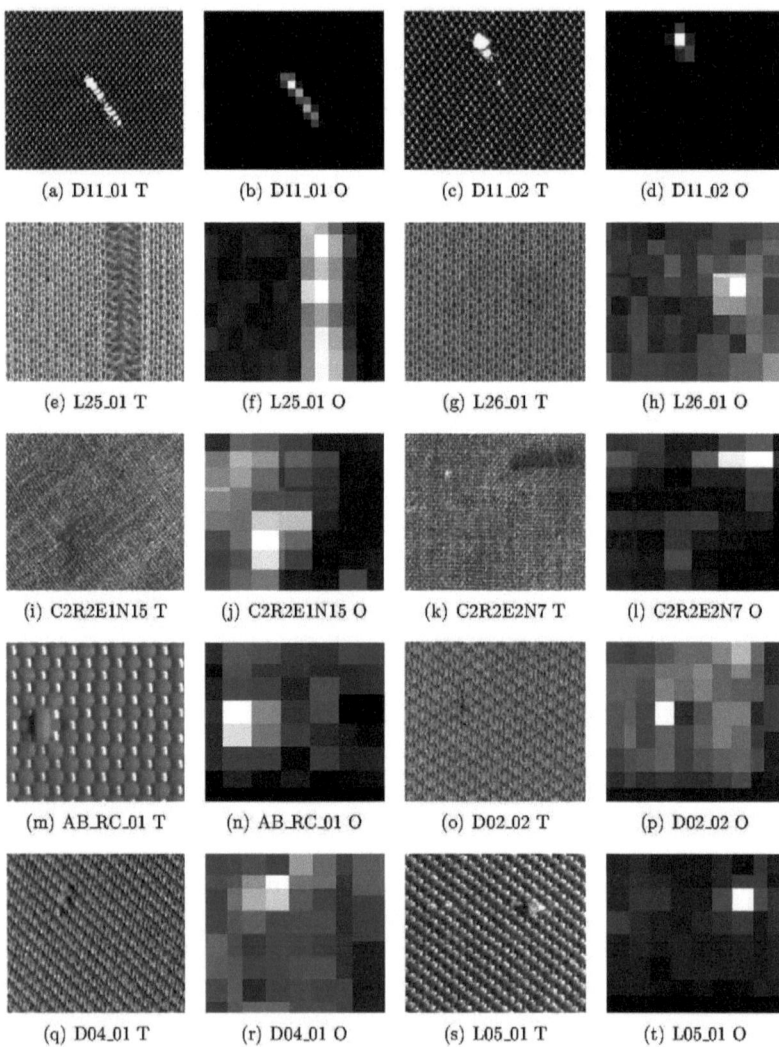

Figure 7.3: Examples of the applicability test on textile defect images. For each example the test image (T) as well as the output (O) of the algorithm is shown. Reference images are not shown. The algorithm output is for illustration purposes not thresholded.

7.3. EXPERIMENTS - RESULTS

Method	PCD (%)	Reference
Decision Fusion	98.64	[TKMA10]
GLCM	97.09	[Mon04]
GLCM + Gabor + wavelet packets (219 features)	96.90	[KFBP08]
Selected from Gabor and GLCM	96.90	[DW01]
Discrepancy Norm Based Template Matching	**96.10**	-
Clustering	91.60	[CS00]
Wavelet97	88.15	[CCC06]
Local Binary Patterns	85.83	[PO99]

Table 7.1: Performance comparison of texture characterization approaches using Percentage of Correct Detection (PCD), for details see Tolba et al. [TKMA10]. The original Table contains multiple entries per reference, here only the best performing ones are listed. Furthermore the top performing method of Murino et al. [MBR04] is skipped because it is a pure classification algorithm without detection. The discrepancy norm based algorithm can be compared with the class of Grey-Level Co-occurrence Matrices (GLCM) and filter (Gabor, wavelet) based feature extraction algorithms.

the most it can its cluster from the origin of the high dimensional space (through the kernel trick) and considers its border as being sustained by few support vectors. Finally the classification is done by taking a linear combination of scalar products of a new input pattern with the supports.

The results on the same set of samples with this method are depicted in Fig. 7.4.

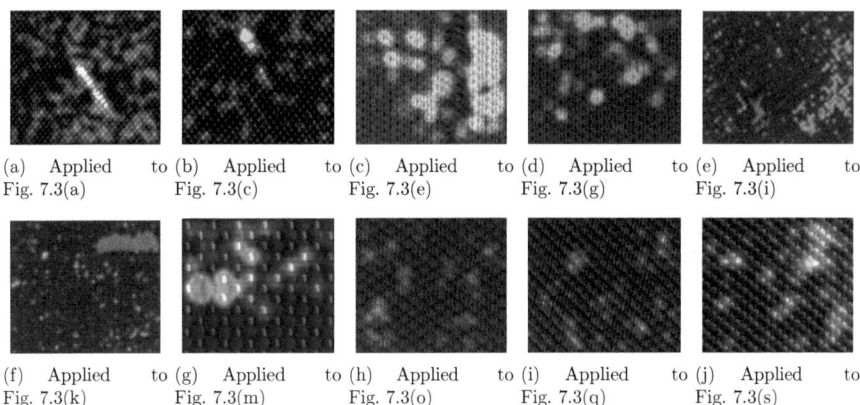

(a) Applied to Fig. 7.3(a) (b) Applied to Fig. 7.3(c) (c) Applied to Fig. 7.3(e) (d) Applied to Fig. 7.3(g) (e) Applied to Fig. 7.3(i)

(f) Applied to Fig. 7.3(k) (g) Applied to Fig. 7.3(m) (h) Applied to Fig. 7.3(o) (i) Applied to Fig. 7.3(q) (j) Applied to Fig. 7.3(s)

Figure 7.4: Results obtained by a OC-SVM applied with a Gaussian kernel directly on the gray-level values.

On this figure, the intensity is adapted to the amount of defectiveness. As we can see, in particular on Fig. 7.4(c) this method is not reliable (though it is also not considered as a competitive state of the art method). The tuning of the parameter is done automatically to optimize the results given a defect-free sample.

Moreover the process is not fast enough for online applications. The images presented here having different sizes (ranging from 150 pixel wide to over 600 pixel high), the algorithm needs sometimes more than a minute to proceed.

Finally, compared to our approach where we have a rather small number of false alarms, the one based on OC-SVM behaves poorly on that point. The smoothness of the outputs should also be considered. For instance, if we notice a defectiveness map continuously growing until a certain maximum then a post processing step could get rid of false alarm (which might be seen as spontaneously appearing in the defectiveness map). However due to the fact the the false alarms are spreading out everywhere on the OC-SVM outputs, such postprocessing steps seem unrealistic compared to the discrepancy norm based case.

Chapter 8

Discussion

8.1 Conclusion

Interested with the structural aspects in image processing and analysis, we have worked on diverse approaches for image representation and interpretation. The first aspects have been driven by the analysis of autocorrelation functions. This should help understanding the important properties that a similarity measure should fulfil in the context of alignment of patterns. As we have seen, it can only be hardly guaranteed to have monotonicity with respect to translation when dealing with pixel-based, or histogram-based similarity measures. Based on a previous work of Moser, the discrepancy norm shows this interesting property which makes it a valuable candidate in the context of alignment. Its practical application has been analysed and some ideas for its approximation by some differentiable functions based on the approximation of the max-norm by some L_p-norms have been introduced. The formula for its derivatives in the continuous case as well as in the discrete case for eventually multidimensional signals are also derived. First results show the robustness of such approximations together with a convergence analysis which bounds the error created.

Along our research on the discrepancy norm, the Hausdorff appears as another good candidate for the analysis of misalignment functions. It also shows the monotonicity property of its autocorrelation function. As a consequence, after reviewing the basics about distance transform for image comparison we introduce a new similarity measure based on the distance transform for sampled functions of Molchanov and Teran. In a last part, motivated by some results from biological vision, we have introduced a new representation which combines binary distance transforms with a scale space description of a signal. Some ideas for image comparison based on this new concept are introduced and a first coarse analysis is given.

Besides these low-level transformation, we have described and analysed different extensions of the analytic signal to higher dimensions. Their mathematical framework based on boundary value problems in higher dimensional complex analysis or Clifford analysis is well described and their computations by means of partial and total Hilbert transforms and Riesz transforms respectively are detailed. We have analysed their practical applications and detailed their features on some typical examples in image processing. Moreover a first analysis on how to conceive a band limited signal and its impact on the computation and the robustness of the computed features is carried over.

The last part of this book was dedicated to some applications of structural image processing tasks and acts also as a motivation for further research in that direction. We have introduced a novel concept in the context of automatic quality control by machine vision for regularly textured surface. In this new paradigm, we make use of a template matching approach to detect defect and make full use of the monotonicity property of the autocorrelation function based on the discrepancy norm. As a result our algorithm shows really promising results and is capable of detecting any kind of defect without cumbersome learning of either the textural component or the particular type of defects one wants to find.

8.2 Opennings

Though many topics have been carried over in this study, there is still some work to be done. This can be described in three aspects.

The first one deals with colour and multispectral images. In this work we have considered only grey-level images and have eventually done a simple linear transformation from the RGB images to a grey value. It would be interesting to have a look at other way of representing images. For this purpose the use of quaternionic and, in a more general way for spectral images, of Clifford algebras might be an interesting path to study. This would allow us to consider the colour information as a single value and no longer as a vector. For instance, in the variational $TV - L_1$ decomposition, the L_1 norm of a colour image is computed by simply integrating the sum of each channel. By using higher dimensional algebra we might describe our image better. This idea has already been used for image representation [BB11]. This colour image analysis can then be applied to each of the proposed approached: discrepancy norm of multichannel images? distance transforms on sampled colour images? Riesz transform on quaternionic representations?

Another aspect is concerned with the use of the approximation proposed for the discrepancy norm. This would allow a better optimisation of the objective function and, hopefully, yield better convergence rate. It would, for instance, avoid the local vanishing derivative which ensures a better convergence.

A last interesting part is about the monogenic representation. A theorem says that the Hilbert transform of a white noise functional is still a white noise, but with some different characteristics, which can be computed from the original one. An interesting point would be to make some inference about the output of the monogenic machinery in order to improve the image analysis part after the Riesz transform. These monogenic representations would also need a deeper understanding in order to ensure the quality of a feature estimation up to a certain bandwidth of the signal. At the moment, this is still an open problem to our knowledge.

References

[AKM95] T. Aach, A. Kaup, and R. Mester. On texture analysis: Local energy transforms versus quadrature filters. *Signal Processing*, 45(2):173–181, 1995.

[AP79] I.E. Abdou and W.K. Pratt. Quantitative design and evaluation of enhancement/thresholding edge detectors. *Proceedings of the IEEE*, 67(5):753–763, 1979.

[AP00] J. Arlandis and J.C. Pérez. Transformation for continuous-valued images. *Pattern recognition and applications*, 56:89, 2000.

[AS66] S.M. Ali and S.D. Silvey. A general class of coefficients of divergence of one distribution from another. *Journal of the Royal Statistical Society. Series B (Methodological)*, pages 131–142, 1966.

[Bad92] A.J. Baddeley. An error metric for binary images. *Robust computer vision*, pages 59–78, 1992.

[Bas96] M. Basseville. Information: entropies, divergences et moyennes (information: entropies, divergences and averages), 1996.

[BB08] L. Bottou and O. Bousquet. The tradeoffs of large scale learning. In J.C. Platt, D. Koller, Y. Singer, and S. Roweis, editors, *Advances in Neural Information Processing Systems*, volume 20, pages 161–168. NIPS Foundation (http://books.nips.cc), 2008.

[BB11] T. Batard and M. Berthier. The spinor representation of images, 2011. Preprint.

[BB13] J.-L. Bouchot and F. Bauer. Discrepancy norm: Approximation and variations. submitted, 2013.

[BBK96] P. Bauer, U. Bodenhofer, and E.P. Klement. A fuzzy algorithm for pixel classification based on the discrepancy norm. In *Fuzzy Systems, 1996., Proceedings of the Fifth IEEE International Conference on*, volume 3, pages 2007–2012. IEEE, 1996.

[BBL02] A. Bodnarova, M. Bennamoun, and S. Latham. Optimal gabor filters for textile flaw detection. *Pattern Recognition*, 35:2973–2991, 2002.

[BBRH13] S. Bernstein, J.-L. Bouchot, M. Reinhardt, and B. Heise. Generalized analytic signals in image processing: Comparison, theory and their applications. In Eckhard Hitzer and Stephen J. Sangwine, editors, *Quaternion and Clifford Fourier Transforms and Wavelets*, Trends in Mathematics. Springer Basel, 2013.

[BC09] J. Beck and W. W. L. Chen. *Irregularities of Distribution*. Cambridge University Press, New York, NY, USA, 2009.

[BDS82] F. Brackx, R. Delanghe, and F. Sommen. Clifford analysis. *Research notes in mathematics*, 76, 1982.

[Bed63] E. Bedrosian. A product theorem for hilbert transforms. *Proceedings of the IEEE*, 51(5):868–869, 1963.

[Bha43] A. Bhattacharyya. On a measure of divergence between two statistical populations defined by their probability distributions. *Bulletin of the Calcutta Mathematical Society*, 35:99–109, 1943.

[BHM10] J.-L. Bouchot, J. Himmelbauer, and B. Moser. On autocorrelation based on hermann weyl's discrepancy norm for time series analysis. In *IJCNN*, pages 1–7. IEEE, 2010.

[BMN13] J.-L. Bouchot and F. Morain-Nicolier. Scaled distance transform and monotonicity of correlations. submitted, 2013.

[BMNMR08] É. Baudrier, F. Morain-Nicolier, G. Millon, and S. Ruan. Binary-image comparison with local-dissimilarity quantification. *Pattern Recognition*, 41(5):1461–1478, 2008.

[BNB04] D. Boukerroui, J.A. Noble, and M. Brady. On the choice of band-pass quadrature filters. *Journal of Mathematical Imaging and Vision*, 21(1):53–80, 2004.

[Bor84] G. Borgefors. Distance transformations in arbitrary dimensions. *Computer vision, graphics, and image processing*, 27(3):321–345, 1984.

[Bor86] G. Borgefors. Distance transformations in digital images. *Computer vision, graphics, and image processing*, 34(3):344–371, 1986.

[BPS00] T. Bülow, D. Pallek, and G. Sommer. Riesz transform for the isotropic estimation of the local phase of moiré interferograms. In *DAGM-Symposium*, pages 333–340, 2000.

[Bro70a] C. G. Broyden. The Convergence of a Class of Double-rank Minimization Algorithms 1. General Considerations. *IMA J Appl Math*, 6(1):76–90, March 1970.

[Bro70b] C. G. Broyden. The Convergence of a Class of Double-rank Minimization Algorithms: 2. The New Algorithm. *IMA J Appl Math*, 6(3):222–231, September 1970.

[BS01] T. Bülow and G. Sommer. Hypercomplex signals-a novel extension of the analytic signal to the multidimensional case. *Signal Processing, IEEE Transactions on*, 49(11):2844–2852, 2001.

[BSM11] J.L. Bouchot, G. Stübl, and B. Moser. A template matching approach based on the discrepancy norm for defect detection on regularly textured surfaces. In *Quality Control by Artificial Vision Conference, QCAV*, 2011.

[BVW12] D. Brunet, E.R. Vrscay, and Z. Wang. On the mathematical properties of the structural similarity index. *Image Processing, IEEE Transactions on*, 21(4):1488–1499, april 2012.

[Can86] J. Canny. A computational approach to edge detection. *Pattern Analysis and Machine Intelligence, IEEE Transactions on*, PAMI-8(6):679–698, 1986.

[Can98] E.J. Candès. *Ridgelets: theory and applications*. PhD thesis, Stanford University, 1998.

[CB01] D. Coquin and P. Bolon. Application of baddeley's distance to dissimilarity measurement between gray scale images. *Pattern Recognition Letters*, 22(14):1483–1502, 2001.

[CB11] M.J. Chen and A.C. Bovik. Fast structural similarity index algorithm. *Journal of Real-Time Image Processing*, 6(4):281–287, 2011.

[CCC06] C.M. Chen, C.C. Chen, and C.C. Chen. A comparison of texture features based on SVM and SOM. In *Proceedings of the 18th International Conference on Pattern Recognition - Volume 02*, ICPR '06, pages 630–633, Washington, DC, USA, 2006. IEEE Computer Society.

[CD00] E.J. Candès and D.L. Donoho. Curvelets: A surprisingly effective nonadaptive representation for objects with edges. Technical report, DTIC Document, 2000.

[CD05a] E.J. Candès and D.L. Donoho. Continuous curvelet transform: I. resolution of the wavefront set. *Applied and Computational Harmonic Analysis*, 19(2):162–197, 2005.

[CD05b] E.J. Candès and D.L. Donoho. Continuous curvelet transform: II. discretization and frames. *Applied and Computational Harmonic Analysis*, 19(2):198–222, 2005.

[Cha01] B. Chazelle. *The discrepancy method: randomness and complexity*. Cambridge Univ Pr, 2001.

REFERENCES

[Cha04] A. Chambolle. An algorithm for total variation minimization and applications. *Journal of Mathematical imaging and vision*, 20(1):89–97, 2004.

[Cha07] S. Chang. Extracting skeletons from distance maps. *IJCSNS*, 7(7):213, 2007.

[CKK04] M. Choi, R.Y. Kim, and M.G. Kim. The curvelet transform for image fusion. *International Society for Photogrammetry and Remote Sensing, ISPRS 2004*, 35:59–64, 2004.

[CKP+07] C. Cadieu, M. Kouh, A. Pasupathy, C.E. Connor, M. Riesenhuber, and T. Poggio. A model of v4 shape selectivity and invariance. *Journal of Neurophysiology*, 98(3):1733–1750, 2007.

[CL97] A. Chambolle and P.L. Lions. Image recovery via total variation minimization and related problems. *Numerische Mathematik*, 76(2):167–188, 1997.

[Cro84] F.C. Crow. Summed-area tables for texture mapping. *ACM SIGGRAPH Computer Graphics*, 18(3):207–212, 1984.

[CS00] H.D. Cheng and Y. Sun. A hierarchical approach to color image segmentation using homogeneity. *IEEE Transactions on Image Processing*, 9(12):2071–2082, 2000.

[Csi63] I. Csiszár. Eine informationstheoretische Ungleichung und ihre anwendung auf den Beweis der ergodizität von Markoffschen Ketten. *Publ. Math. Inst. Hungar. Acad.*, 8:95–108, 1963.

[CST00] N. Cristianini and J. Shawe-Taylor. *An introduction to support Vector Machines: and other kernel-based learning methods*. Cambridge Univ Pr, 2000.

[CV01] T.F. Chan and L.A. Vese. Active contours without edges. *Image Processing, IEEE Transactions on*, 10(2):266–277, 2001.

[CWT+01] Y.T. Chin, H. Wang, L.P. Tay, H. Wang, and W.Y.C. Soh. Vision guided agv using distance transform. In *Proceedings of the 32nd ISR (International Symposium on Robotics)*, volume 19, page 21, 2001.

[CYV00] S.G. Chang, B. Yu, and M. Vetterli. Adaptive wavelet thresholding for image denoising and compression. *Image Processing, IEEE Transactions on*, 9(9):1532–1546, sep 2000.

[CYX06] G.H. Chen, C.L. Yang, and S.L. Xie. Gradient-based structural similarity for image quality assessment. In *Image Processing, 2006 IEEE International Conference on*, pages 2929–2932. IEEE, 2006.

[Dau85] J.G. Daugman. Uncertainty relation for resolution in space, spatial frequency, and orientation optimized by two-dimensional visual cortical filters. *Optical Society of America, Journal, A: Optics and Image Science*, 2:1160–1169, 1985.

[Dau88a] I. Daubechies. Orthonormal bases of compactly supported wavelets. *Communications on pure and applied mathematics*, 41(7):909–996, 1988.

[Dau88b] J.G. Daugman. Complete discrete 2-d gabor transforms by neural networks for image analysis and compression. *Acoustics, Speech and Signal Processing, IEEE Transactions on*, 36(7):1169–1179, 1988.

[Dau90] I. Daubechies. The wavelet transform, time-frequency localization and signal analysis. *Information Theory, IEEE Transactions on*, 36(5):961–1005, sep 1990.

[dBMvGN97] J. F. de Boer, T.E. Milner, M. J. C. van Gemert, and J. S. Nelson. Two-dimensional birefringence imaging in biological tissue by polarization-sensitive optical coherence tomography. *Opt. Lett.*, 22(12):934–936, Jun 1997.

[Der87] R. Deriche. Using canny's criteria to derive a recursively optimal edge detector. *International journal of computer vision*, 1(2):167–187, 1987.

[DFDGtHR04] R. Duits, L. M. J. Florack, J. De Graaf, and B. ter Haar Romeny. On the axioms of scale space theory. *Journal of mathematical imaging and vision*, 20:267–298, 2004.

[Don11] S.-H. Dong. *Wave equation in higher dimensions*. Springer, 2011.

[DT05] N. Dalal and B. Triggs. Histograms of oriented gradients for human detection. In *Computer Vision and Pattern Recognition, 2005. CVPR 2005. IEEE Computer Society Conference on*, volume 1, pages 886–893. Ieee, 2005.

[DW01] A Drimbarean and P F Whelan. Experiments in colour texture analysis. *Pattern Recognition Letters*, 22(10):1161–1167, 2001.

[Dzh96] A. Dzhuraev. On riemann–hilbert boundary problem in several complex variables. *Complex Variables and Elliptic Equations*, 29(4):287–303, 1996.

[FA91] W.T. Freeman and E.H. Adelson. The design and use of steerable filters. *IEEE Transactions on Pattern analysis and machine intelligence*, 13(9):891–906, 1991.

[FB81] M.A. Fischler and R.C. Bolles. Random sample consensus: a paradigm for model fitting with applications to image analysis and automated cartography. *Communications of the ACM*, 24(6):381–395, 1981.

[Fie87] D. J. Field. Relations between the statistics of natural images and the response properties of cortical cells. *J. Opt. Soc. Am. A*, 4(12):2379–2394, Dec 1987.

[Fle70] R. Fletcher. A new approach to variable metric algorithms. *The Computer Journal*, 13(3):317–322, March 1970.

[FS01] M. Felsberg and G. Sommer. The monogenic signal. *Signal Processing, IEEE Transactions on*, 49(12):3136–3144, 2001.

[FS04] M. Felsberg and G. Sommer. The monogenic scale-space: A unifying approach to phase-based image processing in scale-space. *Journal of Mathematical Imaging and vision*, 21(1):5–26, 2004.

[Füh96] H. Führ. Wavelet frames and admissibility in higher dimensions. *Journal of Mathematical Physics*, 37:6353, 1996.

[Gab46] D. Gabor. Theory of communication. part 1: The analysis of information. *Electrical Engineers-Part III: Radio and Communication Engineering, Journal of the Institution of*, 93(26):429–441, 1946.

[Gil06] J. Gilles. Décomposition et détection de structures géométriques en imagerie. Theses de Doctorat, *CMLA ENS Cachan, France*, 2006.

[GO09] T. Goldstein and S. Osher. The split bregman method for l1 regularized problems. *SIAM Journal on Imaging Sciences*, 2(2):323–343, 2009.

[Gol70] D. Goldfarb. A Family of Variable-Metric Methods Derived by Variational Means. *Mathematics of Computation*, 24(109):23–26, January 1970.

[GSZ03] G. Gilboa, N. Sochen, and Y.Y. Zeevi. Texture preserving variational denoising using an adaptive fidelity term. In *Proc. VLsM*, volume 3, 2003.

[GYB04] S.B. Gokturk, H. Yalcin, and C. Bamji. A time-of-flight depth sensor - system description, issues and solutions. In *Computer Vision and Pattern Recognition Workshop, 2004. CVPRW '04. Conference on*, page 35, june 2004.

[Haa10] A. Haar. Zur theorie der orthogonalen funktionensysteme. *Mathematische Annalen*, 69:331–371, 1910.

[Hah92] S.L. Hahn. Multidimensional complex signals with single-orthant spectra. *Proceedings of the IEEE*, 80(8):1287–1300, 1992.

[Hay82] M. Hayes. The reconstruction of a multidimensional sequence from the phase or magnitude of its fourier transform. *Acoustics, Speech and Signal Processing, IEEE Transactions on*, 30(2):140–154, 1982.

REFERENCES

[HHSF92] M.R. Hee, D. Huang, E.A. Swanson, and J.G. Fujimoto. Polarization-sensitive low-coherence reflectometer for birefringence characterization and ranging. *J. Opt. Soc. Am. B*, 9(6):903–908, Jun 1992.

[Hit07] E.M.S. Hitzer. Quaternion fourier transform on quaternion fields and generalizations. *Advances in Applied Clifford Algebras*, 17(3):497–517, 2007.

[HLO80] M. Hayes, J. Lim, and A. Oppenheim. Signal reconstruction from phase or magnitude. *Acoustics, Speech and Signal Processing, IEEE Transactions on*, 28(6):672–680, 1980.

[HLP52] G.H. Hardy, J.E. Littlewood, and G. Pólya. *Inequalities*. Cambridge Univ Pr, 1952.

[HM96] F. Huet and J. Mattioli. A textural analysis by mathematical morphology transformations: Structural opening and top-hat. In *Image Processing, 1996. Proceedings., International Conference on*, volume 3, pages 49–52. IEEE, 1996.

[HM07] A. Haddad and Y. Meyer. An improvement of rudinosher-fatemi model. *Applied and Computational Harmonic Analysis*, 22(3):319–334, 2007.

[HR90] H.J.A.M. Heijmans and C. Ronse. The algebraic basis of mathematical morphology i. dilations and erosions. *Computer Vision, Graphics, and Image Processing*, 50(3):245–295, 1990.

[HSD73] R.M. Haralick, K. Shanmugam, and I. Dinstein. Textural features for image classification. *IEEE Trans. Syst., Man, Cybern*, 3:610–621, 1973.

[HSM+12] B. Heise, S.E. Schausberger, C. Maurer, M. Ritsch-Marte, S. Bernet, and D. Stifter. Enhancing of structures in coherence probe microscopy imaging. In *Proceedings of SPIE*, volume 8335, page 83350G, 2012.

[HSZ87] R.M. Haralick, S.R. Sternberg, and X. Zhuang. Image analysis using mathematical morphology. *Pattern Analysis and Machine Intelligence, IEEE Transactions on*, PAMI-9(4):532–550, 1987.

[IT05] L. Ikonen and P. Toivanen. Shortest routes on varying height surfaces using gray-level distance transforms. *Image and Vision Computing*, 23(2):133–141, 2005.

[JBPN09] S. Jahanbin, A.C. Bovik, E. Pérez, and D. Nair. Automatic inspection of textured surfaces by support vector machines. In *Proceedings of SPIE*, volume 7432, page 74320A, 2009.

[JPS93] D. R. Jones, C. D. Perttunen, and B. E. Stuckman. Lipschitzian optimization without the Lipschitz constant. *Journal of Optimization Theory and Applications*, 79(1):157–181, October 1993.

[KC30] A.N. Kolmogorov and G. Castelnuovo. *Sur la notion de la moyenne*. G. Bardi, tip. della R. Accad. dei Lincei, 1930.

[KFBP08] I. Karoui, R. Fablet, J. M. Boucher, and W. Pieczynski. Fusion of textural statistics using a similarity measure: application to texture recognition and segmentation. *Pattern Analysis and Applications*, 11(3-4):425–434, 2008.

[KN05] L. Kuipers and H. Niederreiter. *Uniform distribution of sequences*. Dover Publications, New York, NY, USA, 2005.

[Koe84] J. Koenderink. The structure of images. *Biological cybernetics*, 50:363–370, 1984.

[Kok35] J.F. Koksma. Ein mengen-theoretischer satz über gleichverteilung modulo eins. *Compositio Math*, 2:250–258, 1935.

[Kov97] P. Kovesi. Symmetry and asymmetry from local phase. *Tenth Australian Joint Converence on Artificial Intelligence*, pages 2–4, 1997.

[Kov99] P. Kovesi. Image features from phase congruency. *Videre: Journal of Computer Vision Research*, 1(3):1–26, 1999.

[Kov02] P. Kovesi. Edges are not just steps. In *Proceedings of the Fifth Asian Conference on Computer Vision*, pages 822–827, 2002.

[Kov03] P. Kovesi. Phase congruency detects corners and edges. School of Computer Science & Software Engineering, The University of Western Australia, 2003.

[Kul97] S. Kullback. *Information theory and statistics*. Dover Pubns, 1997.

[Kum08] A. Kumar. Computer-vision-based fabric defect detection: A survey. *IEEE Trans. Indus. Electr.*, 55:348–363, 2008.

[KYP12] A. Kolamana and O. Yadid-Pecht. Quaternion structural similarity a new quality index for color images. *Image Processing, IEEE Transactions on*, 21(4):1–1, April 2012.

[LBO01] K.G. Larkin, D.J. Bone, and M.A. Oldfield. Natural demodulation of two-dimensional fringe patterns. i. general background of the spiral phase quadrature transform. *JOSA A*, 18(8):1862–1870, 2001.

[LDP07] M. Lustig, D. Donoho, and J.M. Pauly. Sparse mri: The application of compressed sensing for rapid mr imaging. *Magnetic Resonance in Medicine*, 58(6):1182–1195, 2007.

[Lew95] J.P. Lewis. Fast normalized cross-correlation. *Vision Interface*, 10:120–123, 1995.

[LHCH+12] E. Leiss-Holzinger, U.D. Cakmak, B. Heise, J.-L. Bouchot, E.P. Klement, M. Leitner, D. Stifter, and Z. Major. Evaluation of structural change and local strain distribution in polymers comparatively imaged by ffsa and oct techniques. *eXPRESS Polymer Letters*, 6, 2012.

[Lin91] J. Lin. Divergence measures based on the Shannon entropy. *IEEE Transactions on Information Theory*, 37(1):145–151, January 1991.

[Lin94] T. Lindeberg. Scale-space theory: a basic tool for analysig structures at different scales. *Journal of applied statistics*, 21(1-2):225–270, 1994.

[LM70] G. Levi and U. Montanari. A grey-weighted skeleton. *Information and Control*, 17(1):62–91, 1970.

[LMSYAR11] R. Lizarraga-Morales, R. Sanchez-Yanez, and V. Ayala-Ramirez. Homogeneity cues for texel size estimation of periodic and near-periodic textures. *Pattern Recognition*, pages 220–229, 2011.

[Loo53] L.H. Loomis. *An introduction to abstract harmonic analysis*. Van Nostrand Reinhold, 1953.

[Low04] D.G. Lowe. Distinctive image features from scale-invariant keypoints. *International journal of computer vision*, 60(2):91–110, 2004.

[Mal89] S.G. Mallat. A theory for multiresolution signal decomposition: The wavelet representation. *Pattern Analysis and Machine Intelligence, IEEE Transactions on*, 11(7):674–693, 1989.

[Mal99] S.G. Mallat. *A wavelet tour of signal processing*. Academic Pr, 1999.

[MB04] M. Mellor and M. Brady. Non-rigid multimodal image registration using local phase. *Medical Image Computing and Computer-Assisted Intervention–MICCAI 2004*, pages 789–796, 2004.

[MB05] M. Mellor and M. Brady. Phase mutual information as a similarity measure for registration. *Medical Image Analysis*, 9(9):330–343, 2005.

[MBP+08] J. Mairal, F. Bach, J. Ponce, G. Sapiro, and A. Zisserman. Supervised dictionary learning. In *Advances in Neural Information Processing Systems (NIPS)*, 2008.

[MBR04] V. Murino, M. Bicego, and I.A. Rossi. Statistical classification of raw textile defects. In *Proceedings of the Pattern Recognition, 17th International Conference on (ICPR'04)*, volume 4 of *ICPR '04*, pages 311–314, Washington, DC, USA, 2004. IEEE Computer Society.

[Mey01] Y. Meyer. *Oscillating patterns in image processing and non-linear evolution equations: the fifteenth Dean Jacqueline B. Lewis memorial lectures*, volume 22. Amer Mathematical Society, 2001.

[Mey03] Y. Meyer. Modlisation des textures en traitement d'images. Note du sminaire du laboratoire Jacques-Louis Lions, 2003.

[MFB+95] J.-F. Mangin, V. Frouin, I. Bloch, J. Régis, and J López-Krahe. From 3d magnetic resonance images to structural representations of the cortex topography using topology preserving deformations. *Journal of Mathematical Imaging and Vision*, 5:297–318, 1995. 10.1007/BF01250286.

[MFS97] J.L. Marroquin, J.E. Figueroa, and M. Servin. Robust quadrature filters. *JOSA A*, 14(4):779–791, 1997.

[MNLR09] F. Morain-Nicolier, J. Landré, and S. Ruan. Gray level local dissimilarity map and global dissimilarity index for quality of medical images. In *7th IFAC Symposium on Modelling and Control in Biomedical Systems*, pages 281–286, Aalborg, Denmark, aug 2009.

[Mon04] A. Monadjemi. *Towards efficient texture classification and abnormality detection*. PhD thesis, University of Bristol, UK, 2004.

[Mos11] B. Moser. A similarity measure for image and volumetric data based on hermann weyl's discrepancy. *Pattern Analysis and Machine Intelligence, IEEE Transactions on*, 33(11):2321–2329, 2011.

[MS89] D. Mumford and J. Shah. Optimal approximations by piecewise smooth functions and associated variational problems. *Communications on pure and applied mathematics*, 42(5):577–685, 1989.

[MSB11] B. Moser, G. Stübl, and J.-L. Bouchot. On a non-monotonicity effect of similarity measures. *Similarity-Based Pattern Recognition*, pages 46–60, 2011.

[MSRV97] J.L. Marroquin, M. Servin, and R. Rodriguez-Vera. Adaptive quadrature filters and the recovery of phase from fringe pattern images. *JOSA A*, 14(8):1742–1753, 1997.

[MT03] I.S. Molchanov and P. Terán. Distance transforms for real-valued functions. *Journal of mathematical analysis and applications*, 278(2):472–484, 2003.

[NA08] M.S. Nixon and A.S. Aguado. *Feature extraction and image processing*. Academic Press, 2008.

[Nie92] H. Niederreiter. *Quasi-Monte Carlo Methods*. Wiley Online Library, 1992.

[NW87] H. Neunzert and B. Wetton. *Pattern recognition using measure space metrics*. Univ. Kaiserslautern Fachbereich Mathematik, 1987.

[OHCR96] B. Ogor, V. Haese-Coat, and J. Ronsin. Sar image segmentation by mathematical morphology and texture analysis. In *Geoscience and Remote Sensing Symposium, 1996. IGARSS'96. 'Remote Sensing for a Sustainable Future.', International*, volume 1, pages 717–719. IEEE, 1996.

[OL81] A.V. Oppenheim and J.S. Lim. The importance of phase in signals. *Proceedings of the IEEE*, 69(5):529–541, 1981.

[PB01] M. Pesaresi and J.A. Benediktsson. A new approach for the morphological segmentation of high-resolution satellite imagery. *Geoscience and Remote Sensing, IEEE Transactions on*, 39(2):309–320, 2001.

[PM82] T. Peli and D. Malah. A study of edge detection algorithms. *Computer Graphics and Image Processing*, 20(1):1–21, 1982.

[PMV04] J.P.W. Pluim, J.B.A. Maintz, and M.A. Viergever. f-information measures in medical image registration. *Medical Imaging, IEEE Transactions on*, 23(12):1508–1516, 2004.

[PO99] M. Pietikäinen and T. Ojala. Nonparametric texture analysis with simple spatial operator. *Spectrum*, 1999.

[RH91] C. Ronse and H.J.A.M. Heijmans. The algebraic basis of mathematical morphology: Ii. openings and closings. *CVGIP: Image Understanding*, 54(1):74–97, 1991.

[RN02] S. Russell and P. Norvig. *Artificial intelligence: a modern approach (2nd edition)*. Prentice Hall, 2002.

[ROF92] L.I. Rudin, S. Osher, and E. Fatemi. Nonlinear total variation based noise removal algorithms. *Physica D: Nonlinear Phenomena*, 60(1-4):259–268, 1992.

[Rud70] W. Rudin. *Real and Complex Analysis: International Student Edition*. McGraw-Hill, 1970.

[Rud80] W. Rudin. *Function theory in the unit ball of $\mathbb{C}^n$*. Springer, 1980.

[SA90] E.P. Simoncelli and E.H. Adelson. Non-separable extensions of quadrature mirror filters to multiple dimensions. *Proceedings of the IEEE*, 78(4):652–664, 1990.

[SBHM12] G. Stübl, J.-L. Bouchot, P. Haslinger, and B. Moser. Discrepancy norm as fitness function for defect detection on regularly textured surfaces. accepted to DAGM-OAGM, 2012.

[SCD02] J.L. Starck, E.J. Candès, and D.L. Donoho. The curvelet transform for image denoising. *Image Processing, IEEE Transactions on*, 11(6):670–684, 2002.

[SCE01] A. Skodras, C. Christopoulos, and T. Ebrahimi. The jpeg2000 still image compression standard. *Signal Processing Magazine, IEEE*, 18(5):36–58, 2001.

[Sch50] L. Schwartz. *Thorie des distributions*. Hermann, 1950.

[Sch61] L. Schwartz. *Méthodes mathématiques pour les sciences physiques*. Hermann, 1961.

[SCT02] K. Sadakane, N.T. Chebihi, and T. Tokuyama. Discrepancy-based digital halftoning: Automatic evaluation and optimization. In *WTRCV02*, pages 173–198, 002.

[SDZL09] Y. Shi, Y. Ding, R. Zhang, and J. Li. Structure and hue similarity for color image quality assessment. In *Electronic Computer Technology, 2009 International Conference on*, pages 329–333. IEEE, 2009.

[Sha70] D. F. Shanno. Conditioning of Quasi-Newton Methods for Function Minimization. *Mathematics of Computation*, 24(111):647–656, July 1970.

[Shu72] B.O. Shubert. A sequential method seeking the global maximum of a function. *SIAM Journal on Numerical Analysis*, 9(3):379–388, 1972.

[SLHH+11] D. Stifter, E. Leiss-Holzinger, B. Heise, J.-L. Bouchot, Z. Major, M. Pircher, E. Götzinger, B. Baumann, and C.K. Hitzenberger. Spectral domain polarization sensitive optical coherence tomography at 1.55 μm: novel developments and applications for dynamic studies in materials science. *Proceeding of Optical Coherence Tomography and Coherence Domain Optical Methods in Biomedicine XV*, San Francisco, USA 78890Z, 2011.

[SPK95] K.Y. Song, M. Petrou, and J. Kittler. Texture crack detection. *Machine Vision and Applications*, 8:63–76, 1995.

[SS02] B. Schölkopf and A.J. Smola. *Learning with kernels: Support vector machines, regularization, optimization, and beyond*. the MIT Press, 2002.

[STDT08] S. Schuon, C. Theobalt, J. Davis, and S. Thrun. High-quality scanning using time-of-flight depth superresolution. In *Computer Vision and Pattern Recognition Workshops, 2008. CVPRW '08. IEEE Computer Society Conference on*, pages 1–7, june 2008.

[Ste70] E.M. Stein. *Singular integrals and differentiability properties of functions*, volume 30. Princeton Univ Pr, 1970.

REFERENCES

[SWG+09] M.P. Sampat, Z. Wang, S. Gupta, A.C. Bovik, and M.K. Markey. Complex wavelet structural similarity: a new image similarity index. *Image Processing, IEEE Transactions on*, 18(11):2385–2401, 2009.

[SWS+00] B. Schölkopf, R.C. Williamson, A.J. Smola, J. Shawe-Taylor, and J. Platt. Support vector method for novelty detection. *Advances in neural information processing systems*, 12(3):582–588, 2000.

[TDSL00] J.B. Tenenbaum, V. De Silva, and J.C. Langford. A global geometric framework for nonlinear dimensionality reduction. *Science*, 290(5500):2319–2323, 2000.

[tHRF93] B. ter Haar Romeny and L. M. J. Florack. A multiscale geometric model of human vision. In William R. Hendee and Peter N. T. Wells, editors, *Perception of visual information*, pages 73–114. Springer, 1993.

[TKMA10] A.S. Tolba, H.A. Khan, A.M. Mutawa, and S.M. Alsaleem. Decision fusion for visual inspection of textiles. *Textile Research Journal*, 80, 2010.

[TKMB09] F. Timm, S. Klement, T. Martinetz, and E. Barth. Welding inspection using novel specularity features and a one-class svm. In *Proc. of the Int. Conference on Imaging Theory and Applications*, volume 1, pages 146–153, 2009.

[Toi96] P.J. Toivanen. New geodesic distance transforms for gray-scale images. *Pattern Recognition Letters*, 17(5):437–450, 1996.

[Üst10] AS Üstünel. Analysis on wiener space and applications. *Arxiv preprint arXiv:1003.1649*, 2010.

[USVDV09] M. Unser, D. Sage, and D. Van De Ville. Multiresolution monogenic signal analysis using the riesz–laplace wavelet transform. *Image Processing, IEEE Transactions on*, 18(11):2402–2418, 2009.

[Vil48] J. Ville. Theorie et applications de la notion de signal analytique (theory and applications of the notion of analytic signal). *Cables et transmission*, 2(1):61–74, 1948.

[VJ01] P. Viola and M. Jones. Robust real-time object detection. *International Journal of Computer Vision*, 57(2):137–154, 2001.

[VO89] S. Venkatesh and R.A. Owens. An energy feature detection scheme. In *the international conference on image processing*, pages 553–557. Singapore, 1989.

[VV90] P.W. Verbeek and B.J.H. Verwer. Shading from shape, the eikonal equation solved by grey-weighted distance transform. *Pattern Recognition Letters*, 11(10):681–690, 1990.

[VWI97] P. Viola and W.M. Wells III. Alignment by maximization of mutual information. *International journal of computer vision*, 24(2):137–154, 1997.

[WBO97] D.L. Wilson, A.J. Baddeley, and R.A. Owens. A new metric for grey-scale image comparison. *International Journal of Computer Vision*, 24(1):5–17, 1997.

[WBSS04] Z. Wang, A.C. Bovik, H.R. Sheikh, and E.P. Simoncelli. Image quality assessment: From error visibility to structural similarity. *Image Processing, IEEE Transactions on*, 13(4):600–612, 2004.

[Wey16] H. Weyl. Über die gleichverteilung von zahlen mod. eins. *Mathematische Annalen*, 77(3):313–352, 1916.

[WIVA+96] W.M. Wells III, P. Viola, H. Atsumi, S. Nakajima, and R. Kikinis. Multi-modal volume registration by maximization of mutual information. *Medical image analysis*, 1(1):35–51, 1996.

[WSB03] Z. Wang, E.P. Simoncelli, and A.C. Bovik. Multiscale structural similarity for image quality assessment. In *Signals, Systems and Computers, 2003. Conference Record of the Thirty-Seventh Asilomar Conference on*, volume 2, pages 1398–1402. Ieee, 2003.

[Xie08] X. Xie. A review of recent advances in surface defect detection using texture analysis techniques. *Electr. Letters on Computer Vision and Image Analysis*, 3:1–22, 2008.

[You87] R.A. Young. The gaussian derivative model for spatial vision: I. retinal mechanisms. *Spatial Vision*, 2(4):273–293, 1987.

[Zar00] S.K. Zaremba. The mathematical basis of Monte Carlo and Quasi-Monte Carlo methods. *SIAM Review*, 10(3):303–314, 200.

[ZSD05] C. Zhao, W. Shi, and Y. Deng. A new hausdorff distance for image matching. *Pattern Recognition Letters*, 26(5):581–586, 2005.

i want morebooks!

Buy your books fast and straightforward online - at one of world's fastest growing online book stores! Environmentally sound due to Print-on-Demand technologies.

Buy your books online at

www.get-morebooks.com

Kaufen Sie Ihre Bücher schnell und unkompliziert online – auf einer der am schnellsten wachsenden Buchhandelsplattformen weltweit! Dank Print-On-Demand umwelt- und ressourcenschonend produziert.

Bücher schneller online kaufen

www.morebooks.de

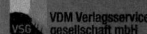

VDM Verlagsservicegesellschaft mbH
Heinrich-Böcking-Str. 6-8   Telefon: +49 681 3720 174   info@vdm-vsg.de
D - 66121 Saarbrücken   Telefax: +49 681 3720 1749   www.vdm-vsg.de

Printed by Books on Demand GmbH, Norderstedt / Germany